Peer-to-peer Accommodation and Community Resilience

Implications for Sustainable Development

Peer-to-peer Accommodation and Community Resilience

Implications for Sustainable Development

Edited by

Anna Farmaki

Dimitri Ioannides

Stella Kladou

CABI

CABI is a trading name of CAB International

CABI
Nosworthy Way
Wallingford
Oxfordshire OX10 8DE
UK

Tel: +44 (0)1491 832111
E-mail: info@cabi.org
Website: www.cabi.org

CABI
WeWork
One Lincoln St
24th Floor
Boston, MA 02111
USA

Tel: +1 (617)682-9015
E-mail: cabi-nao@cabi.org

A catalogue record for this book is available from the British Library, London, UK.

References to Internet websites (URLs) were accurate at the time of writing.

ISBN-13: 9781789246605 (hardback)
9781789246612 (ePDF)
9781789246629 (ePub)

DOI: 10.1079/9781789246605.0000

Commissioning Editor: Claire Parfittt
Editorial Assistant: Lauren Davies
Production Editor: James Bishop

Typeset by SPi, Pondicherry, India
Printed and bound in the UK by Severn, Gloucester

The editors would like to dedicate this book to their children – Michael, Sasha, Ioannis-Alexandros, Alex Ege, Sergios and Alina.

Contents

Biographies of Editors and Authors

Editors

Anna Farmaki is Assistant Professor in Tourism Management at the Cyprus University of Technology, Cyprus. She holds a doctorate (PhD) in marketing. She has worked for several years in private institutions of tertiary education in Cyprus and was course leader of the undergraduate hospitality and tourism management programme at the University of Central Lancashire (Cyprus). She has published extensively in reputable, peer-reviewed academic journals and has presented her work in various international conferences, seminars and workshops. Anna is a Fellow of the Higher Education Academy (UK) and a member of the Management Committee in several COST (European Cooperation in Science and Technology) Actions. She has also been involved in and/or led several nationally and internationally funded research projects. In addition, she is a member of the editorial board of the *International Journal of Contemporary Hospitality Management*, *Tourism Management Perspectives*, *Tourism Review* and the *Journal of Service Theory and Practice*. Her research interests lie primarily in the areas of tourism planning and development with emphasis on sustainable tourism and tourist behaviour.

Dimitri Ioannides is Chaired Professor of Human Geography in the Institution of Economics, Geography, Law and Tourism and a researcher in the European Tourism Research Institute at Mid-Sweden University. Previously, he taught at Missouri State University and held a part-time position at the Centre for Regional and Tourism Research in Bornholm, Denmark. He has published extensively in various journals, including the *Annals of Tourism Research*, *Tourism Management* and the *Journal of Sustainable Tourism*, on issues such as the economic geography of tourism and sustainable development. He is especially interested in matters relating to the social-equity dimension of sustainability, including the geographies of tourism workers and work. He has co-authored books relating to tourism issues, including the *Economic Geography of the Tourist Industry* (Routledge, 1998). Currently, he is an editor of *Tourism Geographies* (in charge of special projects) and serves on the editorial board of several other journals. He also edits the New Directions in Tourism Analysis series (Routledge) as well as serving on the board of the International Polar Tourism Research Network. He is a past recipient of the Roy Wolfe Award for Outstanding Contributions to the Recreation, Tourism and Sport specialty group of the Association of American Geographers.

Stella Kladou is a postdoctoral researcher at the Hellenic Mediterranean University in Crete, Greece. She has served as an assistant professor in the Tourism Administration Department at Boğaziçi University in Istanbul, Turkey and a senior lecturer at Sheffield Hallam University in the UK. Her PhD 'Evaluating cultural city brands: the case of Athens, Istanbul and Rome', was realized in co-operation with the Hellenic Open University in Greece, Istanbul Bilgi University in Turkey and the

University of Rome La Sapienza in Italy, and partially funded by the Italian Ministry of Foreign Affairs and the Scientific and Technological Research Council of Turkey. She has published in prominent academic journals and books of international publishers, and presented at well-esteemed international conferences. She has participated in and/or led several international and national projects as a researcher or consultant. She is an international place brand specialist for The Place Brand Observer, and serves on the scientific board of international conferences and editorial board of the *Journal of Qualitative Research in Tourism*. Her research mainly pertains to place branding, cultural and wine tourism, and digital marketing.

Contributors

Bailey Ashton Adie is currently a visiting research fellow at the Center for Tourism Research, Wakayama University, Japan (previously at Solent University, UK). Her research interests include sustainable tourism, World Heritage tourism, tourism marketing and branding, community resilience to climate change and disasters, second homes, community-based tourism, tourism and development, and heritage tourism. She is the author of the Routledge book, *World Heritage and Tourism: Marketing and Management* (2019). She also sits on the editorial board of the *Journal of Heritage Tourism* and the executive committee of the Leisure Studies Association.

David Albarrán graduated in fine arts at the University of Seville. He did his postgraduate studies in 'Managing tourism and cultural heritage in historic towns' at the University of Malaga. In 2015, he won the *Premio Extraordinario Fin de Estudios* for the most outstanding student on the master's degree in Tourism Management and Planning programme at the University of Seville, where he specialized in public policies. He obtained his PhD degree at the Complutense University, Madrid, with a distinction and international mention. Parallel to his academic career, he has worked as a consultant on tourism and heritage policies for several public organizations in Spain.

Maria del Mar Alonso-Almeida is an assistant professor at the Faculty of Economics and Business Administration at the Autonomous University of Madrid. Her research interests are in circular economy and its applications in destinations, cities and organizations. She has published more than 70 articles in scientific journals. She is also author or co-author of more than ten books. She has participated in 100 conferences both scientific and professional. Her professional experience is in strategic management and change management in organizations. She has also worked for multinational consultancy companies.

Alberto Amore is a lecturer in tourism and air travel management at Solent University, UK. His research interests include urban planning, urban tourism and urban regeneration, destination metagovernance, post-disaster recovery in tourism, destination resilience and tourism and the Sustainable Development Goals. He has published extensively in edited books, peer-reviewed journals, and conference proceedings. He is a Fellow of the Royal Geographical Society and of the Regional Studies Association, and he is a member of the editorial board for *Current Issues in Tourism* and *Tourism Management Perspectives*.

Katerina Antoniou is a lecturer at the University of Central Lancashire, Cyprus, and a course leader for the BA (Hons) in hospitality and tourism management. Dr Antoniou specializes in tourism, peacebuilding and conflict resolution research. Her publications examine patterns of visitation in divided societies, new forms of peace-oriented tourism and the desecuritization of identity. Dr Antoniou holds a BA in political science and economics from Clark University, Massachusetts, an MSc in international relations theory from the London School of Economics, and a PhD in international peacebuilding from the University of Central Lancashire, Preston. Dr Antoniou is a Fulbright Alumna.

Cecilia de Bernardi received her PhD in tourism research from the University of Lapland (Finland) while working at Dalarna University (Sweden). Cecilia is currently a postdoctoral researcher at the European Tourism Research Institute (ETOUR) at Mid-Sweden University. Her publications concentrate mostly on the conceptualization of authenticity in indigenous tourism, especially regarding marketing. She has also published on philosophical issues in academic work and on tourism policy related to winter tourism in the Nordic countries. Cecilia has several research interests related to tourism, but also just pertaining to the philosophical approach critical realism. Her current work focuses on the policy terrain concerning different sport and outdoor activities.

Fani Efthymiadou is a PhD candidate in the Department of Hotel and Tourism Management of Cyprus University of Technology. Her primary research interests are in the areas of the sharing economy, peer-to-peer networks, sustainable development and smart tourism. As a researcher she is involved in several funding programmes.

Olga Gjerald is Associate Professor in Service Management at the Norwegian School of Hotel Management, University of Stavanger, Norway. She holds a PhD in leadership from the University of Stavanger, focusing on basic assumptions of service employees about guests, co-workers and competitors in relation to employee job outcomes. Her main research interests are service organization, management of service employees, and visitor experiences in hospitality and tourism management.

Szilvia Gyimóthy is Associate Professor in Tourism Marketing at Copenhagen Business School. Her research is focused on how tourism is shaping places in the wake of global mobility and mediatized travel. More recently, Szilvia has been working with sharing economy phenomena, and the marketization of social relationships in communitarian businesses.

Chiga Hayakawa holds a BSc from Tokyo University of Science and an MSc in computer science from Queen Mary University of London, UK.

Brian Jones is a senior lecturer at Leeds Business School, Leeds Beckett University. His research strengths, interests and publication outputs range across the areas of corporate social responsibility, enterprise and entrepreneurship education, entrepreneurial marketing, and social media. He has a BA (Hons) degree in sociology from the University of Durham, an MA degree in industrial relations from the University of Warwick and a PhD from the University of Bradford.

Antonios Kaniadakis is a reader in the Department of Computer Science, Brunel University London, UK. He holds a PhD in science, technology and innovation studies from the University of Edinburgh and has published on the social and managerial aspects of digital innovation. He is a senior fellow of the Higher Education Academy.

Christoph Lutz is an associate professor at the Nordic Centre for Internet and Society, BI Norwegian Business School (Oslo). He has a background in sociology (Lic. Phil., University of Zurich), and a PhD in management from the University of St Gallen. Christoph's research interests include digital inequality, privacy, the sharing economy, online participation, and social robots, and he has published widely in top-tier journals such as *New Media & Society*, *Information, Communication & Society*, *Social Media + Society*, *Journal of Management Information Systems*, *Journal of Business Research*, and the *Journal of the Association for Information Science and Technology*.

Filip Majetić is a research associate at the Institute of Social Sciences Ivo Pilar in Zagreb. His main field of research interest is alternative business models (e.g. social entrepreneurship, collaborative economy) and non-standard forms of work (e.g. independent contractors in the gig economy).

Ingeborg Matečić is an assistant professor in the Department of Tourism, Faculty of Economics and Business at the University of Zagreb. She received her first PhD in cultural anthropology from the University of Zagreb, Faculty of Humanities and Social Sciences in 2012. Furthermore, she holds a double doctoral degree (Cotutelle de thèse) in economics with specialization in tourism economics from the University of Quintana Roo, Division of Social Sciences and Administrative Economics and the University of Zagreb, Faculty of Economics and Business. Her research interests lie in the area of cultural tourism and sociocultural sustainability within the concept of sustainable tourism development.

Javier Macías Mendoza has a bachelor's degree in tourism business administration from Eloy Alfaro Ray University of Manabí, Ecuador, and a master's degree in tourism management and consulting from International Iberoamerican University of Mexico. He is currently a PhD candidate at the Rovira i Virgili University in Tarragona, Spain.

Cristina Miguel PhD is Senior Lecturer in digital communication at the University of Gothenburg, Sweden. She received her PhD in media and communications from the University of Leeds, UK.

Her research interests include digital culture (network intimacy, online communities and online privacy) and the digital economy (sharing economy, the political economy of social media and influencer marketing). She has published her research in high-ranked peer-reviewed academic journals, including the *International Journal of Hospitality Management*, the *International Journal of Contemporary Hospitality Management*, *Social Media + Society* and *Convergence*.

Jeroen Oskam (PhD, 1992, Universiteit van Amsterdam) has worked at different universities and hotel and tourism schools in the Netherlands and in Spain. Currently he is the Director of the Research Centre at Hotelschool The Hague (Amsterdam and The Hague, the Netherlands). His own research focuses on the future of hospitality, with recent scenario studies on topics such as online travel agents (OTAs) and urban vacation rentals. His latest books are *Innovation in Hospitality Education* (Springer, 2018), *The Future of Airbnb and the 'Sharing Economy': the Collaborative Consumption of Our Cities* (Channelview, 2019) and *The Overtourism Debate: NIMBY, Nuisance, Commodification* (Emerald, 2020).

Rodrigo Perez-Vega is a senior lecturer in digital marketing and analytics at Kent Business School at the University of Kent. His research interests are in digital marketing, social media and consumer behaviour in online environments, the sharing economy, and applications of artificial intelligence (AI) in marketing. He has published his work in the *Journal of Business Research*, *Tourism Management*, the *Service Industries Journal* and *Marketing Review*. He is also the co-author of the book *Essentials of Digital Marketing* (Goodfellow Publishers, 2018). Rodrigo has also been invited to speak at practitioners' conferences such as the SME (small and medium-sized enterprises) Summit and Social Media Summit. His professional experience is in brand management, entrepreneurship and digital marketing. He has worked for multinationals in fast-moving consumer goods (FMCG) industries and digital marketing agencies.

Michael Röslmaier has a professional degree in hospitality, with a 10 years' international working experience. Thereafter he graduated with a bachelor's degree in tourism studies from the University College Kristiania in Oslo, Norway, with a minor in sociology obtained at the University of California, Berkeley. He obtained his master's degree in human geography at Mid-Sweden University (MIUN), where he remained working as a teaching and research assistant before entering their PhD programme in tourism. He is currently working on his PhD thesis.

Antonio Paolo Russo is Professor of Urban Geography with the Department of Geography, Universitat Rovira i Virgili, Tarragona. He is the coordinator of the PhD programme in tourism and leisure, and lectures in pre-and postgraduate courses in destination management and cultural tourism. Dr Russo is author of 40+ publications in academic journals and books, on research topics ranging from tourism studies to cultural and urban geography. Currently, he leads the H2020 project 'SMARTDEST' tackling tourism mobilities and social exclusion. He is also coordinating the special interest group 'Space Place Mobilities in Tourism' in ATLAS (Association for Tourism and Leisure Education and Research).

Dimitrios Stergiou holds a PhD from the University of Surrey, UK. Currently, he is an assistant professor in tourism management at the School of Social Sciences at Hellenic Open University, Greece. He has over 15 years of experience in teaching, learning and research in the wider areas of tourism education and tourism management. Alongside his academic duties, he has worked with several government and private organizations on tourism project work.

Helena Tolić holds a master's degree in ethnology and cultural anthropology, and Spanish language and literature from the Faculty of Humanities and Social Sciences, University of Zagreb. She worked in the field of heritage interpretation and planning, and cultural management. Her research interests focus on influences of tourism industry on local communities. Her book *The Split Street of Radunica as a Tourist Destination: a Study of Social Interactions* (HED (Hrvatsko etnološko društvo), 2019) is an analysis of the impact of mass tourism on one street in the city of Split in Croatia from the perspective of cultural anthropology.

Jinghua Xie is Associate Professor at both the Norwegian School of Hotel Management, University of Stavanger, and the School of Business and Economics, UiT/The Arctic University of Norway. She gained the Doctor of Philosophy degree specializing in applied economics. Dr Xie has done extensive research on many topics related to consumer demand and market linkages. Her research work on tourism focuses on tourism demand and tourist behaviour.

Acknowledgements

The editors would like to thank a number of people who have played a vital role in the development of this book. First and foremost, we would like to thank all the contributors to this book who have trusted us with their chapters. We appreciate their time, understanding, and patience throughout the numerous revision processes that were undertaken until the publication of the book. We would also like to thank Daniel Guttentag for writing the foreword to this book. Last, we are grateful to the staff at CABI for their patience and support from the preparation of the book proposal to the submission of the final book content. We would especially like to thank Claire Parfitt, Alison Smith and Lauren Davies for their constant encouragement and assistance.

Foreword

When the coronavirus disease 2019 (COVID-19) pandemic began in early 2020, many wondered whether the crisis would derail Airbnb and the broader peer-to-peer short-term rental sector. For example, a headline in *Wired* magazine asked, 'Is this the end of Airbnb?', and a headline in *New York* magazine similarly questioned, 'Will the coronavirus kill Airbnb?'. During this initial stage of the pandemic, the nature of the virus and how it transmitted were still poorly understood, vaccines seemed like a distant fantasy, and global tourism had ground to an abrupt halt. Who – it seemed reasonable to ask – in the foreseeable future would choose to travel somewhere and lodge in a residential home that was owned, managed, and cleaned by a stranger? Likewise, what Airbnb hosts would choose to continue dedicating their properties to tourism lodging when the timetable for guests to return remained so uncertain? In other words, the core practice at the heart of Airbnb's business model – allowing residential homes to be flexibly converted into tourism lodging – was seen by some as a possible source of the company's undoing, as hosts could readily convert those tourist accommodations right back into residential spaces. Moreover, the pandemic crisis was unfolding at what was essentially the worst possible time for Airbnb, right as the company was preparing an initial public offering to launch itself as a public company.

Looking back there is no question that Airbnb was indeed walloped by the pandemic – guests cancelled innumerable reservations, a multitude of hosts exited the platform, and the company laid off nearly 2000 employees. Nevertheless, when one surveys the tourism accommodation landscape today, with the benefit of hindsight, one cannot help but equate the doubts about Airbnb's durability during the pandemic with the doubts about Airbnb's growth potential that many voiced a decade ago, sceptical that it could ever scale to become a meaningful player in the tourism lodging industry that would compete for guests with hotels. Today, it is clear that Airbnb has successfully weathered the pandemic storm and is thriving, having rebounded from the pandemic much faster than most of its hotel counterparts. Owing to travellers' increased desire for privacy (i.e. hygiene and safety), a favourable geographic footprint with concentrations of listings in the types of destinations that travellers have been favouring (e.g. beaches, mountains and lakes), and a natural symbiosis with the rapid rise of remote work, it turned out that Airbnb was particularly well positioned to capitalize on the tourism recovery that has been building since the summer of 2020. In fact, Airbnb even went ahead and launched itself as a public company in December 2020, and presently has a market capitalization over double that of Marriott International Inc. and nearly triple that of Hilton Hotels Corp.

In other words, Airbnb and the broader peer-to-peer short-term rental sector have proven to be extraordinarily resilient, even in the face of the biggest global tourism crisis of our lifetimes. None the less, a far more important question is what Airbnb and the broader peer-to-peer short-term rental phenomenon mean for the resilience of the destinations and host communities in which they operate. This is the critical question that Anna Farmaki, Dimitri Ioannides and Stella Kladou have admirably tackled in the present volume. Airbnb may be successful, but what does that success mean for the success of the communities in which we live and where we visit? Does Airbnb strengthen communities by forging meaningful social connections and generating financial stability for local entrepreneurs, or does Airbnb simply encourage 'touristification' that hollows out the very host communities in which the Airbnb guests supposedly wish to immerse themselves? These questions are as important now as ever before, as the pandemic-induced slowdown in tourism has been seized by numerous destinations as an opportunity to re-evaluate the place that peer-to-peer lodging, and tourism more generally, holds in their communities.

Together, the chapters that make up this volume excel in demonstrating the incredible complexity characterizing the relationship between peer-to-peer lodging and community resilience, by considering multiple stakeholder perspectives and examining a diverse array of destination communities. The research within this book clearly shows how peer-to-peer lodging can foster resilience by helping to make communities more economically, environmentally and socially sustainable. However, the research *also* clearly shows how peer-to-peer lodging can itself be the external force of undesirable change against which communities may struggle to be resilient, and how peer-to-peer lodging can destabilize communities in ways that leave the communities vulnerable and less resilient to future disrupting forces. This conflicting duality highlights the complexity of the peer-to-peer lodging phenomenon, and the nuance with which one therefore must understand it. This complexity can at times be maddening, such as when one wishes to simply boil down the issues to a clear and definitive talking point. None the less, this complexity is also why it is so important to have diverse and rigorous scholarship on the topic, as is found in this volume, and it is additionally why this volume will prove to be so valuable for students, academics, policy makers and community leaders looking to better understand the intersection between peer-to-peer lodging and community resilience.

Daniel Guttentag
College of Charleston
September 2021

Introduction

Dimitri Ioannides[1]*, Stella Kladou[2] and Anna Farmaki[3]
[1]*Mid-Sweden University, Sweden;* [2]*Hellenic Mediterranean University, Greece;*
[3]*Cyprus University of Technology, Cyprus*

Prior to the global outbreak of the COVID-19 pandemic, an article in *The Guardian* (Henley, 2020) lamented the problem of too many visitors that has plagued several European cities, including Amsterdam, Barcelona, Florence and Prague. This piece was one of literally hundreds in recent years, in the popular media but also in academic circles, which address the growing problem of overtourism (Koens *et al.*, 2018; Dodds and Butler, 2019; Milano *et al.*, 2019; UNWTO, 2019; Agyeiwaah, 2020). While dozens of scholars have long acknowledged the issue of overwhelming numbers of tourists in specific areas (including seaside towns and popular cultural venues) (Costa *et al.*, 1996; Mexa and Coccossis, 2004; Van der Borg, 2004), it is the unprecedented magnitude of the problem of overcrowding in an increasing number of places that has alarmed researchers but also policy makers. To make matters worse, it appears that whereas in the past overcrowding predictably occurred in and around a handful of famous sights (e.g. cathedrals, ancient monuments, awe-inspiring natural viewpoints), in more recent years tourists have increasingly been invading mundane spaces such as residential neighbourhoods or so-called neo-bohemian districts (Maitland, 2007; Arias Sans and Quaglieri Domìnguez, 2016; Ioannides *et al.*, 2019; Nieuwland and van Melik, 2020).

The reasons behind this overtourism problem are several and include among others: (i) the emergence of new markets, especially because of the growing number of middle classes in countries such as India and China; (ii) airline deregulation that has made travel, including short-term city breaks, cheaper; (iii) the growing popularity of cruise tourism; and (iv) the appearance of new forms of tourism accommodation associated with so-called peer-to-peer (P2P)-accommodation online platforms, including companies such as Airbnb, HomeAway, 9Flats, HOUSETRIP and Couchsurfing. Other than Couchsurfing that does not work on a fee-based basis (Molz, 2013), these P2P-accommodation platforms allow property owners (hosts) to rent out rooms or entire properties for a short term to visitors. From the perspective of the property provider, the advantage is that through the P2P-accommodation platform they can reach a truly global market whereas from the tourists' viewpoint, they are in a position where they can select from an increasingly diverse range of accommodation options (Guttentag, 2017).

The popularity of P2P accommodation among tourists has surged in recent years. Airbnb, the de facto poster child of this phenomenon (Guttentag, 2019), has grown impressively since its inception in 2007 into a company that,

*Dimitri.Ioannides@miun.se

as of 30 September 2020, boasted 5.6 million listings in more than 100,000 cities worldwide. At that time, there were more than 4 million hosts and average earnings per host were US$9600/year (Airbnb, n.d.). An important outcome of the P2P-accommodation phenomenon in numerous localities around the world is that entire neighbourhoods have often been transformed into short-term rental areas, especially when landlords recognize the opportunity to maximize their profits by renting out to visitors as opposed to long-term residents. Simply said, it is far more lucrative to rent an apartment on Airbnb in a city like Athens for €150 or more per night as opposed to renting it out long term for a monthly fee of €600. This, of course, has meant that in popular parts of many cities, property values have skyrocketed, leading to a crisis in the availability of affordable housing and growing citizen anger (Füller and Michel, 2014; Colomb and Novy, 2017; Oppilard, 2017).

Notably, in 2015, several demonstrators invaded the Airbnb headquarters in San Francisco to express their support for a citywide bill to increase controls on the short-term rental market. Their anger reflected that of many citizens throughout the city who had become increasingly alarmed about landlords terminating tenants' contracts and refusing to rent to people on lower incomes such as students in order to profit from much higher rents they could obtain from P2P-accommodation guests (Booth, 2015). Similarly, an article in *The Guardian* (Hickey and Cookney, 2016) described the growing opposition in several cities against P2P-accommodation platforms, aiming to introduce regulations to curb the phenomenon. For instance, Barcelona had introduced a rule whereby persons wanting to rent their homes through an online P2P-accommodation platform had to apply for a licence while Reykjavik imposed time limits as to how long one could rent out their property before being obliged to be taxed as a business. Meanwhile, in the Athens neighbourhood of Exarchia, left-wing activists began a campaign to push out Airbnbs, which they regarded as the cause of widespread gentrification (Bateman, 2019). In this area, which is close to the city centre and has a history of radicalism, there have been accusations that because of P2P accommodation, many long-term residents have been evicted.

While much debate exists as to whether P2P accommodation causes gentrification or whether, in fact, gentrification leads property owners to turn to P2P accommodation as a way to supplement their mortgage payments (Arias Sans and Quaglieri Domìnguez, 2016; Füller and Michel, 2014), important questions arise. These include: (i) What does this phenomenon actually mean in terms of community sustainability?; (ii) How do residents, but also existing businesses, in localities that are inundated by P2P-accommodation activity cope and adapt to this phenomenon?; (iii) In what manner does the influx of P2P accommodation transform certain areas?; and (iv) Are some neighbourhoods more amenable to such activities than others? Further, we should also investigate what P2P accommodation means for different types of communities. Does it, for instance, have differing repercussions for places in the countryside as opposed to urban regions? Finally, given that at the time of putting the finishing touches to this book we are still experiencing the COVID-19 pandemic that began in early 2020, we should ask what this global crisis means for destinations in the future and the way they will cope with disruptive innovations, including P2P accommodation. These are some of the issues that this edited collection seeks to flesh out. In the rest of this introductory chapter, we provide a brief review of the relevant literature on the topic of P2P accommodation and what this means for community sustainability and resilience. We end by presenting the outline of the book's contents.

The P2P-Accommodation Phenomenon

Given the growing popularity of P2P accommodation and the effects it has caused on local communities and economies, it is unsurprising that many researchers have investigated several aspects relating to the phenomenon. The bulk of the studies draw from the Airbnb case, reflecting the dominance of this particular platform in the market over other competitors. Guttentag (2019, p. 815) maintains that 'although Airbnb is far from the only company dedicated to peer-to-peer short-term rentals, owing to its immense size and vast impacts, most would agree that

Airbnb is the most important company of its type'. This, most likely, explains why most studies on the P2P-accommodation phenomenon appear to focus on Airbnb rather than its competitors.

There have been several attempts in recent years to review pertinent literature on P2P accommodation as it relates to short-term rentals and tourism specifically (e.g. Prayag and Ozanne, 2018; Belarmino and Koh, 2020; Sainaghi, 2020). It is also worth mentioning Guttentag's (2019) comprehensive literature review relating specifically to Airbnb. This study alone identified 132 peer-reviewed English-language articles on the topic published in a variety of journals, although the most popular outlets were journals relating to tourism and hospitality. These studies have identified key research areas at the micro, meso and macro levels.

To begin with, there are studies which seek to conceptualize P2P-accommodation forms, distinguishing these from traditional ones (Huber, 2017; Farmaki *et al.*, 2020b). While commercial home settings such as bed and breakfasts (B&Bs) and guesthouses have been around for a long time (Sheringham and Daruwalla, 2007), the rise of the collaborative economy (Dredge and Gyimóthy, 2017) facilitated by the growth of Internet-mediated markets that enable individuals to gain access to products and services for a fee (Gupta *et al.*, 2019) has led to the development of an alternative accommodation offering, namely P2P accommodation. In P2P accommodation, there is a service provider (such as Airbnb) that acts as an intermediary between the supplier (i.e. the host) and the customer (i.e. the guest) who will pay for underutilized products and/or services (Kumar *et al.*, 2018). Hence, unlike other forms of commercial home settings, P2P accommodation creates value by allowing underutilized assets to become accessible online to individuals (Stephany, 2015). For instance, through the P2P-accommodation platform, one can rent out unused bedrooms in one's home or alternatively can list an entire house (e.g. a second home) for short-term rental when the owner does not need it.

Many studies focusing on the drivers of P2P-accommodation exchanges have identified several benefits associated with the sector. These represent the second stream of research on P2P accommodation. This body of work reveals both economic and social benefits, which drive both guests and hosts to participate in P2P-accommodation exchanges. The most predominant of these are value for money, entrepreneurial opportunities, additional income, convenience, novelty, socialization opportunities and perceived authenticity (Tussyadiah and Pesonen, 2018; Farmaki and Stergiou, 2019; Lin *et al.*, 2019; Farmaki and Kaniadakis, 2020). Unlike in the case of commercial accommodation settings (e.g. hotels), in P2P accommodation the social interactions between hosts and guests and the 'home feeling' offered are considered central to the P2P-accommodation experience (Lyu *et al.*, 2019; Moon *et al.*, 2019; Zhu *et al.*, 2019).

Yet another line of research in the literature delves into user-centric investigations revolving around trust issues (e.g. Ert and Fleischer, 2019; Guttentag, 2019; Sthapit and Björk, 2019), as well as guest and host behaviours, experiences and satisfaction (e.g. Liang *et al.*, 2018; Giannoni *et al.*, 2021). These studies highlight the importance of platform features and policies in driving demand and maximizing trust between hosts, guests and the platform whereas host types, practices and pricing strategies have also been examined, shedding light on host–guest roles and relationships (Boto-García *et al.*, 2021; Xie *et al.*, 2021).

Notwithstanding, a significant research stream of P2P-accommodation literature that has attracted the attention of scholars and policy makers alike is devoted to the examination of impacts of P2P-accommodation growth. The focus of these studies is twofold. First, there is research examining the effects of P2P accommodation on the tourism sector itself, including businesses such as hotels. Certain studies compare traditional and P2P accommodation across a range of characteristics such as location, pricing and so on (e.g. Zervas *et al.*, 2017; Dogru *et al.*, 2020; Sainaghi and Baggio, 2020; Li *et al.*, 2021). Zervas *et al.* (2017), for example, discovered that Airbnbs led to lower revenues especially in the case of lower-end properties. Conversely, research has tended to be less conclusive in the case of high-end hotels (Guttentag and Smith, 2017; Zervas *et al.*, 2017). Guttentag (2019) points out the existence of studies where hotel-industry representatives do not seem too bothered about P2P accommodation since they feel that this caters to a different type of clientele than their own. Despite this obvious uncertainty as to the real effect of P2P accommodation on more traditional tourism establishments, there does seem to be a general agreement by representatives of the latter sector that companies such as Airbnb should be better regulated (Koh and King, 2017).

The second type of research concerning P2P-accommodation effects relates to studies focusing specifically on how the growth of platforms such as Airbnb have affected local economies and communities at large. There are several investigations, which explore the characteristics of the P2P-accommodation units. For example, some authors examine the market share of company-owned Airbnb properties, where one entity controls several properties, in destinations such as Reykjavik (Adamiak, 2018; Ioannides *et al.*, 2019). Adamiak (2018) offers a comprehensive overview of the characteristics but also the distribution of Airbnbs throughout Europe, and identifies that the number of such properties in each locality corresponds to their population size but also whether these destinations are tourist oriented (e.g. historic cities, coastal or mountain resorts). Another finding is that in some cities, especially in Southern Europe, there is more capacity in Airbnbs as opposed to hotels. Adamiak also points out where in Europe the Airbnb phenomenon is more likely to be professionally owned (many properties owned by one entity) as opposed to locations where they are likelier to be individually owned. For instance, in the Nordic countries, properties are usually owned by individuals, whereas in certain Mediterranean countries the properties are run by professional companies.

It has also become increasingly popular to study the geographical distribution of P2P-accommodation units, especially in metropolitan areas. In his comprehensive literature review of Airbnb, Guttentag (2019) highlights several studies that examine in which parts of cities there is a tendency to find a high concentration of these short-term rentals, especially in central areas. Indeed, as Guttentag (2019, p. 829) states 'this body of research has largely concluded that Airbnb listings reflect fairly similar spatial patterns to hotels' and that in many cities they tend to be located within close proximity to various attractions. There are, of course, exceptions. Ioannides *et al.* (2019) concluded from their study in Utrecht that certain types of neighbourhoods, which they describe as neo-bohemian, are also likely to develop into secondary Airbnb hubs. These areas are ones with a good mix of restaurants, bars and cafés and offer a vibrant scene of markets, independent movie theatres and ethnic shops, all of which are attractive to students, immigrants and also members of the creative class (Maitland, 2007). They are located not too far from the central business districts of cities or in neighbourhoods that are well connected to public transportation (Wegmann and Jiao, 2017).

Another significant research track when it comes to the effects of P2P accommodation on various communities are concerns relating to what the expansion of these short-term rentals means for the supply of affordable housing for locals. As we have already mentioned in the opening section of this Introduction, there is a debate as to whether the proliferation of properties in certain neighbourhoods causes gentrification (Wachsmuth and Weisler, 2017). In a study of Los Angeles, Lee (2016) noted the fact that short-term rentals caused a greater shortage of affordable housing. Lee's argument is that once a home is listed on a platform such as Airbnb this means that it is removed from the supply of properties for long-term residents. Lee believes that it is natural for homeowners to want to maximize their profits and so once they realize how lucrative it is to make money from P2P rentals, they have less incentive to rent out to long-term residents who are on fixed incomes. In a study in New Orleans, Johnson (2015) noted how the character of several areas in the city has drastically been transformed because of the rapid dispersal of Airbnb rentals. Whether or not P2P rentals are a direct cause of gentrification is an issue that Füller and Michel (2014) examined with reference to the Kreuzberg district of Berlin. In this case, they argue that gentrification had already been caused through the growing demand of many people to buy properties that were previously rentals in the district. Many of these homes were bought as investments and several serve the purpose of being second homes. Many owners have realized the improved potential for generating a profit through renting these out through Airbnb or similar platforms.

Because of the growing alarm concerning the effects of P2P accommodation in several places in the world, there are plenty of cases where authorities have begun to aggressively seek to regulate these properties (Crommelin *et al.*, 2018; Guttentag, 2019). In a study of several US and European cities as well as a detailed case study of Denver, Nieuwland and van Melik (2020) observe that, in most cases, regulations are imposed in order to mitigate the effects of short-term rentals on the price of housing and on the character of neighbourhoods. These authors argue, however, that despite the existence of regulatory instruments in several cities their implementation is difficult to enforce. They conclude that clearly 'many cities are far from figuring out how to handle this new player in the tourism field' (Nieuwland and van Melik, 2020, p. 823). Theirs and other studies (e.g. Tham, 2016; Wegmann and Jiao, 2017) also note that the types of regulations aimed at curtailing the effects of short-term rentals vary considerably from country to country and from locality to locality. This often has to do with why the regulation was deemed necessary in the first place. As Niewland and van Melik (2020) state, cities like Barcelona or Amsterdam which have been inundated with Airbnbs and other short-term rentals have reacted to concerns about overcrowding and the shortage of affordable housing for locals. Meanwhile, Denver has adopted more of a 'let's see what happens' attitude, contending that it is hard to have a one-size-fits-all regulation since the effects vary from neighbourhood to neighbourhood.

P2P Accommodation, Sustainability and Community Resilience

As seen in the preceding paragraphs, the explosion of P2P accommodation has drastically transformed the tourism playing field in numerous municipalities. On the one hand, companies like Airbnb provide benefits to hosts such as the ability to earn extra money from using spare assets (e.g. empty bedrooms or second homes). They also offer tourists the option to expand their selection considerably when looking for lodgings at their destinations. On the other hand, the rapid explosion of this sector has led to several negative effects on various places. This raises the question as to what the advent of P2P accommodation means in terms of community sustainability and overall resilience. Surprisingly, despite the enormous literature concerning sustainability and tourism and the emerging studies linking community resilience to the sector (e.g. Cheer and Lew, 2018), there have not been many attempts to explore the interlinkages of sustainability, resilience and P2P accommodation.

A cursory review of the literature has revealed few papers that collectively use terms such as P2P and sustainability or P2P and resilience or, even, Airbnb and community resilience. Using geographic information systems (GIS), Garcia-Ayllon (2018) points out that the social and economic sustainability of various Spanish cities has been seriously compromised because of the largely speculative activity associated with the real estate market that is exclusively aimed at further developing the P2P sector. Álvarez-Herranz and Macedo-Ruíz (2021) also explore how Airbnb rentals have influenced neighbourhood sustainability in a quantitative study of Madrid for the period 2015–2018. They note that in certain neighbourhoods of the Spanish capital, Airbnb has contributed to overtourism, which in turn has significantly reduced the supply of housing while the economic, social and environmental sustainability of these areas has been seriously compromised. A further quantitative study of the effects of short-term rentals on the heritage resources of Turin in Italy demonstrated that Airbnbs in historical residential areas can have a particularly adverse impact on the built environment (Rubino *et al.*, 2020). This leads to calls for the development of indicators for measuring the socio-environmental effect of short-term rentals.

When it comes to resilience in the context of P2P accommodation, it appears that we have barely broken the surface in terms of research. A rare examination of where destination resilience is linked to Airbnb development is provided by Volgger *et al.* (2018). Resilience, as defined by these authors, has to do with how a community adapts and transforms when its tourism system is disrupted. This relates not only to coping with the changes that have occurred but also to how the community reacts and even innovates (Volgger *et al.*, 2018). This study relied on interviews with a few stakeholders in Western Australia to gauge their perceptions of Airbnb's effects but also the opportunities it may create in their communities. They also explored what the P2P-accommodation phenomenon means in terms of social capital formation. They concluded that 'the Airbnb phenomenon may challenge and affect a destination's resilience in various and, in particular, in mixed ways' (Volgger *et al.*, 2018, p. 148). A particularly thorny problem in terms of gauging the resilience of communities in the context of tourism is that the sector is highly fragmented (Dredge and Gyimothy, 2017). Being able to successfully respond to disruptions that relate to events such as the advent of P2P accommodation has much to do with how effectively and closely the various actors involved in a destination's governance can collaborate.

Thus, this edited book aims to shed light on the interrelationships between sustainability, resilience and P2P accommodation. Representing a collection of 11 chapters written by 25 contributors, this book attempts to contribute to the ongoing discussion on the impacts of the growing P2P accommodation phenomenon, highlighting the important role of local communities' resilience for the sustainable development of destinations impacted by the growth of P2P-accommodation platforms. Through a collection of empirical and conceptual chapters, this book may be informative to academics and policy makers alike involved in tourism, but also in other disciplines, who may be interested in the sustainable development of communities. Below, an overview of the structure of the edited book is provided.

Structure of the Edited Book

The book begins with an exploration of the impacts of P2P accommodation in relation to community resilience at the neighbourhood level. In Chapter 1, Mendoza and Russo explore the issue of 'airbnbization' in urban spaces by focusing on Barcelona, a metropolitan area that has 'suffered' heavily (prior to the COVID-19 outbreak) because of forces leading to gentrification and the emergence of overtourism (Milano *et al.*, 2019). The focus is primarily on the Vila de Gràcia neighbourhood. Here, inflation of rents and housing prices overall have compromised housing affordability and this situation has, in turn, transformed the neighbourhood. Mendosa and Russo also delve into other factors, which have resulted from the proliferation of short-term rentals that have pushed away many local residents from the area. Drawing from secondary data and residents' personal stories, the authors portray how the case of Vila de Gràcia in Barcelona offers a canvas of new avenues of tourism-driven gentrification overlapping traditional ones. Vila de Gràcia originally escaped – at least partially – from excessive tourism pressure in terms of overcrowding, littering, impoverishment and banalization of the commercial offer. Yet, the neighbourhood became 'the place to be', thereby recently revealing an increasing process of population substitution and internationalization rather than displacement and shrinkage. Residents, old and new, acknowledge the benefits of the visitor economy, but are concerned with commercial change and conviviality issues in residential spaces.

Chapter 2 investigates the case of the Koukaki neighbourhood in Athens, Greece. Stergiou, Farmaki and Efthymiadou highlight the tourism impacts that occur at the level of neighbourhoods and delve further into residents' perspectives on the impacts of P2P accommodation. They reveal a dominance of negative perceptions relating to socio-economic and environmental impacts emerging from the influx of Airbnb tourists in residential areas. The authors develop a framework of residents' perceptions of P2P-accommodation impacts, which illustrates, among others, how these perceptions lead to residents' reactions and more passive or active forms of behaviour on their behalf to adapt to the new conditions that are caused due to P2P-accommodation growth in their neighbourhoods. This work draws from social exchange theory to identify residents' attempts to mitigate associated impacts and highlight the importance of the mediating role of platforms and property owners in the social exchanges of residents and guests and, ultimately, in the resilience of the local community.

The following chapter (Chapter 3) by Kaniadakis, Hayakawa and Farmaki shifts the focus from Europe to the Far East and Japan in particular. Rather than focusing on neighbourhoods, this chapter looks into P2P-accommodation as a solution to the shortage of accommodation facilities in relation to anticipated increasing tourist numbers, especially in light of the planned 2020 Tokyo Olympic and Paralympic Games (which were eventually moved to 2021 due to the COVID-19 pandemic). Well-known events such as the Olympic and Paralympic Games traditionally lead to enhanced numbers of temporary visitors in the designated venue city. These include not only spectators but also large numbers of, among others, athletes, journalists, officials and people who are employed as volunteers or regular staff (Delaplace and Schut, 2019). Kaniadakis, Hayakawa and Farmaki acknowledge the role of the Private Lodging Business Act, which was established in Japan in 2018 in an attempt to regulate P2P accommodation and which led to most listings being removed from the Airbnb platform. Building on Japanese Airbnb hosts' perceptions, this chapter presents findings relevant to both urban and rural areas, as it embraces listings that include the hosts' own homes as well as second homes not regularly used. The authors first illustrate how individual, non-professional hosts constitute a source of community resilience. Yet, due to the Act, such hosts have now been marginalized by the platform. This professionalization of hosting in Japan is an example of the shift of P2P accommodation to a more profit-centric model. The chapter concludes with implications on how this shift limits the potential for meaningful visitor–host contact, which is key for community resilience. It also emphasizes the importance of non-professional hosts seeking to formalize the resilient role of the Airbnb platform as well as their own resilience through lobbying, knowledge-sharing and self-organizing activities (Berkes and Ross, 2013).

Tolić and Matečić in Chapter 4 provide a longitudinal perspective of short-term rentals in the neighbourhood of the Croatian city of Split. They discuss how, over time, the process of operating short-term rentals in this city has shifted. They use the term 'offline P2P' accommodation, which was adopted as a post-crisis recovery tool in the aftermath of the war during the 1990s. They focus specifically on one street, Radunica Street, situated close to the city centre to identify the original motives for starting offline P2P accommodation and resilient hosts' attributes. Rather than the dyadic host–guest relationship witnessed in many other destinations, they recognize in this case how the so-called 'gatherer–host–tourist' cooperation has changed the culture of everyday life (gatherers are locals who search for tourists). Initially, the post-war socio-economic situation in Croatia prompted many individuals to engage, first with paid offline and eventually online P2P accommodation. Going beyond the 'host-to-tourist' way of doing business helped the local community develop community resilience, in three phases. Once the local community accepted the new circumstances, they adopted an affirmative look towards P2P accommodation and, finally, started cooperating and developed community networks, collaborations and relationships.

In Chapter 5, Miguel, Lutz, del Mar Alonso-Almeida, Jones, Majetić and Perez-Vega employ a holistic approach to resilience by exploring sociocultural, economic, political, environmental and technological impacts of short-term rentals in the UK context. They provide some insights concerning the impacts of P2P accommodation as perceived by platform users, which relate mostly to environmental care, urban transformation, social capital and lifestyle, housing, infrastructure and events, and antisocial behaviour and crime. Overall, the respondents in this study perceive the impact of short-term rentals as positive. Discussing the findings from a community resilience angle, the authors suggest that the increase in P2P guest figures can contribute to positive changes in cities (Alonso-Almeida *et al.*, 2019) if policy makers introduce efficient regulation of the short-term rental sector aimed at drastically reducing its negative impacts and enhancing its positive ones. Such a situation can lead to more favourable perceptions while improving community resilience by building on community identity and lifestyle.

In Chapter 6, Xie and Gjerald primarily focus on the environmental impacts of P2P accommodation and specifically guests' pro-environmental behaviour. Their primary objective relates to exploring the importance of guests' pro-environmental behaviour as a means of improving host communities' resilience and for mitigating negative impacts associated with the growth of P2P accommodation. Drawing on a Norwegian sample, this study provides an insightful first understanding on how Airbnb guests' environmental knowledge and concern significantly influence their behaviour. The chapter sheds light on how attracting such guests and designing relevant strategies to promote their pro-environmental behaviour can benefit destination governance and improve community resilience.

Chapters 7 and 8 move the focus to rural areas. Specifically, in Chapter 7 Adie, de Bernardi and Amore offer a comparative case study of two established second-home destinations, namely north Wales and Dalarna in Sweden. Both these regions have witnessed a proliferation of P2P accommodation because of their popularity as attractive rural settings. In both regions, the way primary residents, second homeowners and Airbnb users perceive rural space mediates the spatial interaction in these rural environments. Although Airbnb guests' perceptions of rurality are similar to those of other rural tourists, Adie *et al.* highlight the influence of local hosts and the wider rural community. Compared to locals, Airbnb users prioritize proximity to amenities and attractions and generally discuss different dimensions of rurality. The authors suggest that a deeper understanding of how different users of the same rural space perceive rurality is critical for resilience planning.

Röslmaier and Albarrán in Chapter 8 prioritize a sense-of-place perspective when it comes to how P2P accommodation transforms rural settings. A mixed-methods approach helps the authors explore a key rural destination, that of the island of Heimaey in Iceland. Because of how Airbnb developments influence sense of place and community resilience, residents in areas where there are high concentrations of 'professional' Airbnbs (i.e. ones owned by companies who often control more than one property) feel a sense of nostalgia about the 'lost past'. Consequently, they often display a strong intention to withdraw from these places. Conversely, locals in more isolated areas and landscapes, where non-professional Airbnb properties exist, maintain their place identity and indicate a strong sense of place and loyalty to their communities. Chapter 8, therefore, highlights how sense of place can both serve and hinder community resilience depending on the place dynamics of Airbnb.

Whereas most chapters in this edited book examine P2P-accommodation platforms that involve paid transactions (such as Airbnb), Chapter 9 explores a different type of P2P accommodation for its potential as a peacebuilding tool on the divided island of Cyprus. Here, Antoniou focuses on the Cypriot Couchsurfing community, which includes members from both the Greek Cypriot and the Turkish Cypriot communities. She approaches community resilience in relation to sociocultural and political perspectives. In her chapter, community resilience achieved through the Couchsurfing platform is not an end goal. Rather, it is valued for its contribution to the creation of strong bonds across members of divided societies and as the basis of sustainable peace.

Resilience can, in fact, be a positive outcome not only at the state level, but, as Gyimóthy in Chapter 10 suggests, in neighbourhood communities as well. Her study emphasizes that a tradition of communitarian movements in Nordic countries (including Denmark and Sweden) has meant that the negative impacts observed elsewhere because of the often uncontrolled growth of P2P-accommodation platforms, have not materialized to the same extent. For instance, the structure of homeownership is such that it is extremely hard for one entity to control many residential properties at once as has been the case in, for example, Greece. Delving specifically into the evolution and cultural history of Danish housing cooperative movements, Gyimóthy argues her case concerning the value of participatory democracy and cohabitation and their respective contribution to building resilience.

Chapter 11 is the last chapter of this edited volume which uses two scenario studies conducted in 2015 and 2018 to help communities in strategic decision making regarding tourism development. Through this analysis, Oskam contributes to the assessment of the long-term impacts of P2P accommodation on resident communities while looking back at his 'home-sharing' scenarios to see which of the underlying assumptions have shown to be accurate and which chain of events has actually evolved. This evaluation can show to what extent scenarios have been useful in supporting destination resilience, which can be even more beneficial taken the currently changing patterns in P2P accommodation and the sudden decrease of international and urban tourism since 2020 due to the COVID-19 pandemic, and the contribution of scenario planning in anticipating and managing disruptions.

Concluding Remarks

At the time of the writing of this book, the world was experiencing an unprecedented situation.

The COVID-19 pandemic, which was initially detected in Wuhan (China) in December 2019 (Yang *et al.*, 2020), had spread quickly across the world. By July 2021, a total of 3.98 million deaths and 184 million cases of infected people were reported worldwide (WHO, 2021). In an attempt to contain the diffusion of the virus, national governments imposed local lockdowns, travel restrictions and even closed their borders, bringing international travel to a standstill. As a result, the global tourism industry was severely hit with the aviation and hospitality sectors experiencing tremendous economic losses (Farmaki *et al.*, 2020a; Ioannides and Gyimothy, 2020). Airlines suspended their operations, hotels and restaurants shut down while staff were laid off or made redundant (Alonso *et al.*, 2020; Sahin, 2020). By the beginning of 2021, vaccinations with high effectiveness against the pandemic had been developed; however, their uneven distribution worldwide, not to mention the appearance of new waves and variants of the coronavirus, highlighted the pandemic's endurance.

In this context of uncertainty, media reports suggested that the COVID-19 pandemic had brought the P2P-accommodation sector to its knees. The main rationale behind this argument is twofold. On the one hand, the P2P-accommodation sector experienced decreasing supply, since a large number of hosts – especially non-professional ones – made their properties unavailable and/or turned them into long-term rents (Farmaki *et al.*, 2020a). On the other hand, P2P accommodation experienced decreasing demand. This was largely because of lack of standardization in host practices, which meant that the potential guests were unsure both about the hygiene conditions in the properties and the terms of cancellation (Glusac, 2020). Further, fears of contamination and social-distancing guidelines led to the minimization of interaction between hosts and guests, thus undermining one of the most important drivers of the P2P-accommodation experience (De Canio *et al.*, 2020).

Indeed, the sector has experienced the crippling effects of the pandemic with platforms such as Airbnb reporting millions of losses in revenue due to the rising number of cancellations and lack of booking requests (Farmaki *et al.*, 2020a). In this regard, P2P-accommodation platforms were heavily criticized for reflecting a lack of strategic thinking in response to the pandemic (Carpenter, 2020). For example, Airbnb announced at the beginning of the COVID-19 outbreak a refund policy, which led to the anger of guests when they found out that some hosts offered partial refunds or travel credit (Webster, 2020). Likewise, hosts expressed their frustration over the lack of financial support they received from governments which regarded the sector as 'non-essential business' (Evans, 2020). Consequently, platforms announced their intentions to support the host community. For instance, Airbnb claimed it would provide US$250 million to its host community and set up a relief fund for superhosts. (Superhost status is characterized by a high rate of honouring reservations, a high percentage of five-star reviews, a prompt and high response rate and longer hosting experience, clearly indicating high standards of service provision.) Similarly, Couchsurfing turned from a free to a paid P2P-accommodation platform during the COVID-19 pandemic, asking members to contribute a monthly fee to keep the community alive.

Despite these difficulties, first indications suggest that the sector will survive, albeit with adaptations to improve platform resilience. For instance, Farmaki *et al.* (2020a) identified a range of responses that P2P-accommodation hosts undertake to deal with the pandemic, including adaptation of their practices in line with traditional accommodation standards. The authors concluded that such measures may be necessary in the near future to maximize guests' level of confidence in the sector in terms of safety and hygiene. Alternatively, Zhang *et al.* (2021) concluded that hosts that are innovative and embrace the original features of the P2P exchange will boost the P2P-accommodation sector in the post-COVID-19 era and eliminate those who dilute the uniqueness of the sector. Indeed, the pandemic has uncovered numerous opportunities that may be leveraged by hosts. However, a large proportion of non-professional hosts in particular are thinking about exiting from P2P-accommodation platforms and turning to longer-term renting instead (Farmaki *et al.*, 2020a). While this possibility may at first appear as a panacea to the problems local communities face due to P2P-accommodation growth (i.e. the housing crisis), it may also signify the domination of professional hosts in the sector.

Thus, important questions arise over the future evolution of the P2P-accommodation sector in light of a potential changing profile of its host community and the implications of this on destinations' sustainable development and local communities' resilience. Evidently, more attention should be paid on the nexus between P2P accommodation, sustainability and local community resilience. This edited volume represents a small contribution to the important discourse surrounding this topic by presenting a range of empirical and conceptual perspectives from urban and rural communities. Each chapter discusses a specific topic that may impart significant insights to policy makers, practitioners and academicians alike. Although the collection of the chapters presented in this edited volume does not cover all of the potential impacts of P2P accommodation's growth on local communities and/or local communities' approaches in dealing with negative impacts imposed by the phenomenon's rapid expansion, significant viewpoints are presented that make a contribution to the ongoing discussion on P2P accommodation. We hope this book provides a stepping stone for further exploration of key issues pertinent to the interface between P2P accommodation, local community resilience and destination sustainable development and, as such, provide ideas for more effective destination management and inclusive destination governance.

References

Adamiak, C. (2018) Mapping Airbnb supply in European cities. *Annals of Tourism Research* 71, 67–71.

Agyeiwaah, E. (2020) Over-tourism and sustainable consumption of resources through sharing: the role of government. *International Journal of Tourism Cities* 6(1), 99–116.

Airbnb (n.d.) About us. Available at: https://news.airbnb.com/about-us/ (accessed 3 August 2021).

Alonso, A.D., Kok, S.K., Bressan, A., O'Shea, M., Sakellarios, N., *et al.* (2020) COVID-19, aftermath, impacts, and hospitality firms: an international perspective. *International Journal of Hospitality Management* 91: 102654.

Alonso-Almeida, M.M., Borrajo-Millán, F. and Yi, L. (2019) Are social media data pushing overtourism? The case of Barcelona and Chinese tourists. *Sustainability* 11(12), 3356.

Álvarez-Herranz, A. and Macedo-Ruíz, E. (2021) An evaluation of the three pillars of sustainability in cities with high Airbnb presence: a case study of the city of Madrid. *Sustainability* 13, 3220.

Arias Sans, A. and Quaglieri Domìnguez, A. (2016) Unravelling Airbnb: urban perspectives from Barcelona. In: Russo, A.P. and Richards, G. (eds) *Reinventing the Local in Tourism: Producing, Consuming and Negotiating Place*. Channel View Publications, Bristol, UK, pp. 209–228.

Bateman, J. (2019) Exarchia: How Airbnb angered Greek anarchists. BBC News, 19 October. Available at: https://www.bbc.com/news/blogs-trending-50075503 (accessed 19 October 2019).

Belarmino, A. and Koh, Y. (2020) A critical review of research regarding peer-to-peer accommodations. *International Journal of Hospitality Management* 84: 102315.

Berkes, F. and Ross, H. (2013) Community resilience: toward an integrated approach. *Society & Natural Resources* 26(1), 5–20.

Booth, K. (2015) Protesters occupy Airbnb HQ ahead of housing affordability vote. *The Guardian*, 2 November. Available at: https://www.theguardian.com/us-news/2015/nov/02/airbnb-san-francisco-headquarters-occupied-housing-protesters (accessed 2 November 2015).

Boto-García, D., Mayor, M. and De la Vega, P. (2021) Spatial price mimicking on Airbnb: multi-host vs single-host. *Tourism Management* 87: 104365.

Carpenter, A. (2020) Airbnb demonstrates lack of strategic thinking in response to COVID-19. *VRMintel*, 6 May. Available at: https://vrmintel.com/airbnb-demonstrates-lack-of-strategic-thinking-in-response-to-covid-19 (accessed 8 May 2020).

Cheer, J. and Lew, A. (eds) (2018) *Tourism, Resilience and Sustainability: Adapting to Social, Political and Economic Change.* Routledge, London.

Colomb, C. and Novy, J. (eds) (2017) *Protest and Resistance in the Tourist City*. Routledge, London.

Costa, P., Gotti, C. and Van der Borg, J. (1996) Tourism in European heritage cities. *Annals of Tourism Research* 23(2), 306–321.

Crommelin, L., Troy, L. Martin, C. and Pettit, C. (2018) Is Airbnb a sharing economy superstar? Evidence from five global cities. *Urban Policy and Research* 36(4), 429–444.

De Canio, F., Nieto-Garcia, M., Martinelli, E. and Pellegrini, D. (2020) The motives behind consumers' intention to use peer-to-peer accommodation: an fsQCA application. *International Journal of Contemporary Hospitality Management* 32(9), 2969–2989.

Delaplace, M. and Schut, P.O. (eds) (2019) *Hosting the Olympic Games: Uncertainty, Debates and Controversy*. Routledge, London.

Dodds, R. and Butler, R. (2019) *Overtourism: Issues, Realities and Solutions*. Walter de Gruyter, Berlin.

Dogru, T., Hanks, L., Mody, M., Suess, C. and Sirakaya-Turk, E. (2020) The effects of Airbnb on hotel performance: evidence from cities beyond the United States. *Tourism Management* 79: 104090.

Dredge, D. and Gyimóthy, S. (eds) (2017) *Collaborative Economy and Tourism: Perspectives, Politics, Policies and Prospects*. Springer, Cham, Switzerland.

Ert, E. and Fleischer, A. (2019) The evolution of trust in Airbnb: a case of home rental. *Annals of Tourism Research* 75, 279–287.

Evans, D. (2020) Airbnb's future is uncertain as it continues to struggle through its Covid-19 response. *CNBC*, 6 May. Available at: https://www.cnbc.com/2020/05/06/can-airbnb-survive-the-coronavirus-pandemic.html (accessed 11 May 2020).

Farmaki, A. and Kaniadakis, A. (2020) Power dynamics in peer-to-peer accommodation: insights from Airbnb hosts. *International Journal of Hospitality Management* 89: 102571.

Farmaki, A. and Stergiou, D.P. (2019) Escaping loneliness through Airbnb host–guest interactions. *Tourism Management* 74, 331–333.

Farmaki, A., Miguel, C., Drotarova, M.H., Aleksić, A., Časni, A.Č. and Efthymiadou, F. (2020a) Impacts of Covid-19 on peer-to-peer accommodation platforms: host perceptions and responses. *International Journal of Hospitality Management* 91: 102663.

Farmaki, A., Stergiou, D.P. and Christou, P. (2020b) Sharing economy: peer-to-peer accommodation as a foucauldian heterotopia. *Tourism Review* 76(3), 570–578.

Füller, H. and Michel, B. (2014) Stop being a tourist! New dynamics of urban tourism in Berlin-Kreuzberg. *International Journal of Urban and Regional Research* 38(4), 1304–1318.

Garcia-Ayllon, S. (2018) Urban transformations as an indicator of unsustainability in the P2P mass tourism phenomenon: the Airbnb case in Spain through three case studies. *Sustainability* 10, 2933.

Giannoni, S., Brunstein, D., Guéniot, F. and Jouve, J. (2021) Multichannel distribution strategy of Airbnb hosts. *Annals of Tourism Research Empirical Insights* 2(1): 100017.

Glusac, E. (2020) Hotels vs. Airbnb: Has Covid-19 disrupted the disrupter? *The New York Times*. Available at: https://www.nytimes.com/2020/05/14/travel/hotels-versus-airbnb-pandemic.html (accessed 1 November 2021).

Gupta, M., Esmaeilzadeh, P., Uz, I. and Tennant, V.M. (2019) The effects of national cultural values on individuals' intention to participate in peer-to-peer sharing economy. *Journal of Business Research* 97, 20–29.

Guttentag, D. (2017) Regulating innovation in the collaborative economy: an examination of Airbnb's early legal issues. In: Dredge, D. and Gyimóthy, S. (eds) *Collaborative Economy and Tourism: Perspectives, Politics, Policies and Prospects*. Springer, Cham, Switzerland, pp. 97–128.

Guttentag, D. (2019) Progress on Airbnb: a literature review. *Journal of Hospitality and Tourism Technology* 10(4), 814–844.

Guttentag, D.A. and Smith, S.L. (2017) Assessing Airbnb as a disruptive innovation relative to hotels: substitution and comparative performance expectations. *International Journal of Hospitality Management* 64, 1–10.

Henley, J. (2020) Overtourism in Europe's historic cities sparks backlash. *The Guardian*, 25 January. Available at: https://www.theguardian.com/world/2020/jan/25/overtourism-in-europe-historic-cities-sparks-backlash (accessed 25 January 2020).

Hickey, S. and Cookney, F. (2016) Airbnb faces worldwide opposition. It plans a movement to rise up in its defence. *The Guardian*, 29 October. Available at: https://www.theguardian.com/technology/2016/oct/29/airbnb-backlash-customers-fight-back-london (accessed 29 October 2016).

Huber, A. (2017) Theorising the dynamics of collaborative consumption practices: a comparison of peer-to-peer accommodation and cohousing. *Environmental Innovation and Societal Transitions* 23, 53–69.

Ioannides, D. and Gyimothy, S. (2020) The COVID-19 crisis as an opportunity for escaping the unsustainable global tourism path. *Tourism Geographies* 22(3), 624–632.

Ioannides, D., Röslmaier, M. and van der Zee, E. (2019) Airbnb as an instigator of 'tourism bubble' expansion in Utrecht's Lombok neighbourhood. *Tourism Geographies* 21, 822–840.

Johnson, C. (2015) Gentrifying New Orleans: thoughts on race and the movement of capital. *Souls* 17(3–4), 175–200.

Koens, K., Postma, A. and Papp, B. (2018) Is overtourism overused? Understanding the impact of tourism in a city context. *Sustainability* 10(12), 4384.
Koh, E. and King, B. (2017) Accommodating the sharing revolution: a qualitative evaluation of the impact of Airbnb on Singapore's budget hotels. *Tourism Recreation Research* 42(4), 409–421.
Kumar, V., Lahiri, A. and Dogan, O.B. (2018) A strategic framework for a profitable business model in the sharing economy. *Industrial Marketing Management* 69, 147–160.
Lee, D. (2016) How Airbnb short-term rentals exacerbate Los Angeles' affordable housing crisis: analysis and policy recommendations. *Harvard Law & Policy Review* 10, 229–253.
Li, J., Hudson, S. and So, K.K.F. (2021) Hedonic consumption pathway vs. acquisition-transaction utility pathway: an empirical comparison of Airbnb and hotels. *International Journal of Hospitality Management* 94: 102844.
Liang, L.J., Choi, H.C. and Joppe, M. (2018) Exploring the relationship between satisfaction, trust and switching intention, repurchase intention in the context of Airbnb. *International Journal of Hospitality Management* 69, 41–48.
Lin, P.M., Fan, D.X., Zhang, H.Q. and Lau, C. (2019) Spend less and experience more: understanding tourists' social contact in the Airbnb context. *International Journal of Hospitality Management* 83, 65–73.
Lyu, J., Li, M. and Law, R. (2019) Experiencing P2P accommodations: anecdotes from Chinese customers. *International Journal of Hospitality Management* 77, 323–332.
Maitland, R. (2007) Tourists, the creative class and distinctive areas in major cities: the roles of visitors and residents in developing new tourism areas. In: Richards, G. and Wilson, J. (eds) *Tourism, Creativity and Development*. Routledge, New York, pp. 73–86.
Mexa, A. and Coccossis, H. (2004) Tourism carrying capacity: a theoretical overview. In: Coccossis, H. and Mexa, A. (eds) *The Challenge of Tourism Carrying Capacity Assessment: Theory and Practice*. Ashgate, Farnham, UK.
Milano, C., Cheer, J. and Novelli, M. (2019) *Overtourism: Excesses, Discontents and Measures in Travel and Tourism*. CAB International, Wallingford, UK.
Molz, J. (2013) Social networking technologies and the moral economy of alternative tourism: the case of couchsurfing.org. *Annals of Tourism Research* 43, 210–230.
Moon, H., Miao, L., Hanks, L. and Line, N.D. (2019) Peer-to-peer interactions: perspectives of Airbnb guests and hosts. *International Journal of Hospitality Management* 77, 405–414.
Nieuwland, S. and van Melik, R. (2020) Regulating Airbnb: how cities deal with perceived negative externalities of short-term rentals. *Current Issues in Tourism* 23(7), 811–825.
Oppilard, F. (2017) From San Fransisco's 'Tech boom 2.0' to Valparaiso's UNESCO World Heritage Site: resistance to tourism gentrification from a comparative political perspective. In: Colomb, C. and Novy, J. (eds) *Protest and Resistance in the Tourist City*. Routledge, London, pp. 129–151.
Prayag, G. and Ozanne, L.K. (2018) A systematic review of peer-to-peer (P2P) accommodation sharing research from 2010 to 2016: progress and prospects from the multi-level perspective. *Journal of Hospitality Marketing & Management* 27(6), 649–678.
Rubino, I., Coscia, C. and Curto, R. (2020) Identifying spatial relationships between built heritage resources and short-term rentals before the Covid-19 pandemic: exploratory perspectives on sustainability issues. *Sustainability* 12, 4533.
Sahin, T. (2020) Lufthansa applies for short-time work assistance. Available at: https://www.aa.com.tr/en/economy/lufthansa-applies-for-short-time-work-assistance/1788026 (accessed 15 April 2020).
Sainaghi, R. (2020) The current state of academic research into peer-to-peer accommodation platforms. *International Journal of Hospitality Management* 89: 102555.
Sainaghi, R. and Baggio, R. (2020) Substitution threat between Airbnb and hotels: myth or reality? *Annals of Tourism Research* 83: 102959.
Sheringham, C. and Daruwalla, P. (2007) Transgressing hospitality: polarities and disordered relationships? In: Lashley, C., Lynch, P. and Morrison, A. (eds) *Hospitality: a Social Lens*. Elsevier, Oxford, pp. 33–46.
Stephany, A. (2015) *The Business of Sharing: Making it in the New Sharing Economy*. Springer, Cham, Switzerland.
Sthapit, E. and Björk, P. (2019) Sources of distrust: Airbnb guests' perspectives. *Tourism Management Perspectives* 31, 245–253.
Tham, A. (2016) When Harry met Sally: different approaches towards uber and Airbnb – an Australian and Singapore perspective. *Information Technology and Tourism* 16(4), 393–412.

Tussyadiah, I.P. and Pesonen, J. (2018) Drivers and barriers of peer-to-peer accommodation stay – an exploratory study with American and Finnish travellers. *Current Issues in Tourism* 21(6), 703–720.
UNWTO (World Tourism Organization) (2019) *Overtourism? Understanding and Managing Urban Tourism Growth beyond Perceptions. Volume 2: Case Studies*. UNWTO, Madrid.
Van der Borg, J. (2004) Tourism management and carrying capacity in heritage cities. In: Coccosis, H. and Mexa, A. (eds) *Challenges of Tourism Carrying Capacity Assessment*. Ashgate, Avebury, UK, pp. 163–179.
Volgger, M., Pforr, C. and Reiser, D. (2018) Destination resilience and the sharing economy. In: Innerhofer, E., Fontanari, M. and Pechlaner, H. (eds) *Destination Resilience: Challenges and Opportunities for Destination Management and Governance*. Routledge, London, pp. 142–152.
Wachsmuth, D. and Weisler, A. (2017) Airbnb and the rent gap: gentrification through the sharing economy. *Environment and Planning A: Economy and Space* 50(6), 1147–1170.
Webster, L. (2020) Covid-19: Booking.com suspends UK bookings after pressure from MPs. *The National.* Available at: https://www.thenational.scot/news/18394049.covid-19-booking-com-suspends-uk-bookings-pressure-mps/ (accessed 12 May 2020).
Wegmann, J. and Jiao, J. (2017) Taming Airbnb: towards guiding principles for local regulation of urban vacation rentals based on empirical results from US cities. *Land Use Policy* 79, 494–501.
WHO (World Health Organization) (2021) Coronavirus Disease (COVID-19) Situation Report – 71. Available at: https://www.who.int/docs/default-source/coronaviruse/situation-reports/20200331-sitrep-71-covid-19.pdf?sfvrsn=4360e92b_6 (accessed 2 July 2021).
Xie, K., Heo, C.Y. and Mao, Z.E. (2021) Do professional hosts matter? Evidence from multi-listing and full-time hosts in Airbnb. *Journal of Hospitality and Tourism Management* 47, 413–421.
Yang, Y., Zhang, H. and Chen, X. (2020) Coronavirus pandemic and tourism: dynamic stochastic general equilibrium modeling of infectious disease outbreak. *Annals of Tourism Research* 83: 102913.
Zervas, G., Proserpio, D. and Byers, J.W. (2017) The rise of the sharing economy: estimating the impact of Airbnb on the hotel industry. *Journal of Marketing Research* 54(5), 687–705.
Zhang, M., Geng, R., Huang, Y. and Ren, S. (2021) Terminator or accelerator? Lessons from the peer-to-peer accommodation hosts in China in responses to COVID-19. *International Journal of Hospitality Management* 92: 102760.
Zhu, Y., Cheng, M., Wang, J., Ma, L. and Jiang, R. (2019) The construction of home feeling by Airbnb guests in the sharing economy: a semantics perspective. *Annals of Tourism Research* 75, 308–321.

1 Mobile Gentrifiers and Leavers: Tourist Dwelling as an Agent of Exclusion in Barcelona

Javier Macías Mendoza[1]* and Antonio Paolo Russo[2]
[1]*University Rovira i Virgili, Spain;* [2]*University Rovira i Virgili, Spain*

Abstract
The rise in popularity of short-term rentals (STRs) has been widely noted to affect housing affordability, thus presenting itself as a driver of social exclusion in tourist cities. It is also suggested, but less documented, that rent inflation is not the only factor that may be pushing autochthonous residents out of neighbourhoods experiencing intense 'airbnbization'. Dwelling in STRs unfolds in ways that could be challenging to many aspects of everyday life for a stable population, such as night rest, security, familiarity with neighbours, health, and access to basic services. Our research on such issues is carried out in Vila de Gràcia, a neighbourhood of Barcelona where traditional gentrification is enmeshed with the rising tourist profile of the area. In our fieldwork we examined how residents perceive and cope with such pressures and investigated how (and to what extent) the material hindrances produced by temporary dwelling can trigger the decision to move out.

Keywords: short-term rentals, everyday life, social change, social conflict, Barcelona, dwelling

Highlights

- This chapter contributes to the debate on the expansion of short-term rentals, focusing on community impacts at neighbourhood level.
- The case of Vila de Gràcia in Barcelona offers a canvas of new avenues of tourism-driven gentrification overlapping traditional ones.
- In this area, the reconversion of housing uses to short-term rentals may be contributing to increased hardships for the retention of long-term residents.
- Residents, old and new, are aware of the benefits of the visitor economy, but wary of commercial change and conviviality issues in residential spaces.

Introduction

This chapter intends to contribute to the debate on the new avenues of social exclusion that have been unfolding in tourist places in recent years, namely housing exclusion, in relation to the penetration of platform-mediated accommodation. We focus specifically on the mechanisms by which the growth of 'home-stays' as short-term rentals (STRs) becomes a source of disruption of the social and population fabric of neighbourhoods.

*javier27mm@hotmail.com

(eds Anna Farmaki, Dimitri Ioannides and Stella Kladou)
DOI: 10.1079/9781789246605.0001

In a burgeoning literature, it is quite well documented that STRs bring about a restructuring of the real estate market. The inflation of rents and housing prices pushes the weakest segments of the population out of the areas of high concentration of tourist rentals (Dredge *et al.*, 2016; Segú, 2018; Wachsmuth and Weisler, 2018; Lagonigro *et al.*, 2020). Observers have highlighted that this outcome is substantially driven by rent extraction tactics: (i) proprietors make higher profits and enjoy a more flexible regime of 'temporary availability' of their property when they rent houses or rooms in the short-term market (A. Quaglieri-Domínguez, A. Arias Sans and A.P. Russo, 2021, unpublished data); or (ii) 'buy to let' (Cócola-Gant and Gago, 2021), which is emerging as a form of 'spatial fix' driving the investment strategies of corporate speculators and sectors of the middle classes (Semi and Tonetta, 2020).

Some authors have noted that other neighbourhood transformations are enmeshed with these processes. First is the inflow of transnational 'nomads' pushing gentrification. These populations, characterized by mobile biographies and high levels of cognitive capital, select areas offering amenities also sought by urban tourists and increasingly catering to cosmopolitan urbanites and 'transient' consumers, often at the expense of the needs and affordability of the incumbent 'sedentary' population (Füller and Michel, 2014; Jover and Díaz-Parra, 2019; López-Gay *et al.*, 2020). Second is the 'signposting' of certain neighbourhoods as the new speculative frontier in tourist cities, with a fundamental role of the platforms themselves in providing guidance and technical support to area narratives and even home interior design (Bialski, 2016). Rent differentials are thus sustained by symbolic capital, responding to the market opportunities opened by contemporary 'post-tourist' practices (Hiernaux and González, 2014).

Cócola Gant (2016) observes that the displacement of original residents from touristified neighbourhoods takes different simultaneous and interrelated forms. The first is direct displacement that occurs with the expulsion of tenants when hotels and tourist apartments are installed in residential spaces. The second refers to displacement by exclusion when the rental or purchase of a home becomes unattainable. The third is the result of difficult coexistence with tourists, affecting the daily life of the residents and leading them to leave the neighbourhood. Finally, the fourth refers to collective displacement, whereby the social transformation of an area prevents the reproduction of residential life. While in-depth studies of the first two processes have followed (González-Pérez, 2019; Ardura Urquiaga *et al.*, 2020; Cócola Gant and López-Gay, 2020; Petruzzi *et al.*, 2020), little attention has appeared in the literature about the narratives and biographies of the population at risk of exclusion in STR-ridden neighbourhoods. One of the few exceptions is A. Quaglieri-Domínguez *et al.* (2021, unpublished data), who studied the life courses and biographies of the host community established in El Raval, an area of Barcelona characterized by wide ethnic and class heterogeneity. Even before they are actually displaced by the housing market dynamics, these populations may be affected in different ways by the social and cultural change experienced in their livelihood through: (i) marginalization; (ii) erosion of their local 'support systems'; (iii) disruption of daily routines and rest; and (iv) changes in feelings of security and safety. These processes – possibly constituting (or leading to) emerging forms of social exclusion in a tourist context – are becoming an important element of critical stances towards tourism growth in urban contexts (Novy, 2019).

Recent research on overtourism and its effects (Milano *et al.*, 2019; Salerno and Russo, 2020) has given new life and sense to traditional analytic concepts such as host attitudes towards visitors, as in Doxey's (1975) evolutionary model leading to antagonism. Yet we still miss a more solid analytic approach to make sense of the interdependencies between elements of social pressure brought about by tourism growth, shifting sensibilities on tourism, and the dynamics of constituencies (displacement, polarization, social and political change). These issues have been explored in Russo and Scarnato (2018) as part of an evolutionary approach to the production of critical discourses with tourism growth, and, in a similar vein, by Richards *et al.* (2019). The topic of population displacement also touches upon the debate on place resilience. In this sense, we follow a part of the literature that conceives place resilience as a normative and subversive concept (Grove, 2014; Vale, 2014), hinting at the capacity of space to guarantee social reproduction and social justice when exposed to the agency of processes of place transformation such as tourism. Thus, resilient destinations are not those which have a 'built-in' capacity to bounce back in terms of performance when subject to external shocks, or to adapt to breakthrough innovations (such as the digitalization of the hospitality marketplace and the related boom of STR), but rather those that in the face of these changes are able to maintain the well-being and inclusion of the community through appropriate policy responses. This ability may well be one of the greatest current challenges for cities in the face of touristification processes (Sequera and Nofre, 2018).

Through our work, we intend to offer a contribution to this regard, exploring in some depth how and to what degree the hindrances brought about by an increasing penetration of STR at neighbourhood level affects certain population groups, and how that may lead them to decide to move out. Towards this end, we carry out a study in a neighbourhood of Barcelona, Vila de Gràcia, which we consider a representative area of the processes under scrutiny.

The chapter is structured as follows. The first section provides the broader canvas of the research on tourism and population change in Vila de Gràcia. We illustrate the trajectory of Barcelona as a successful example of urban regeneration towards being a world-class destination and its eventual 'overtourism' crisis. With this backdrop in mind, we situate Vila de Gràcia as a neighbourhood where processes of population substitution are highly enmeshed with its rising profile as an unconventional tourist area characterized by a high level of penetration of STR. In the following section we present our research on community perceptions of tourists dwelling in Vila de Gràcia. This section exploits secondary sources (the municipal survey of residents' perception on tourism) as well as the information obtained through a survey and some in-depth interviews with a sample of former and current residents; its aim is to reconstruct how different types of residents react to or cope with the increasing numbers of tourists dwelling in their neighbourhood. Finally, in the last section, we discuss the implications of our findings in the light of the existing literature on neighbourhood change, touristification and resilience.

Background: Barcelona Between Urban Restructuring and 'Airbnbization'

Tourism and social exclusion in Barcelona

Several researchers have explored the issues of tourism's social impacts in Barcelona. This could be explained by the fact that the Catalan city has experienced a rapid transformation into a global destination since its 'Olympic period' at the end of the 1980s, against the backdrop of a social fabric, which at the end of the Francoist period and during the transition to democracy was still characterized by strong polarization (Monclús, 2003; Smith, 2005). The city's newly gained positioning as an attractive hub for people and capital in the post-industrial global economy has been noted as a successful pioneering effort, bringing together public and private interests in a pervasive programme of urban regeneration. However, at the same time, it has opened new avenues of social inequality that only became apparent after the 'Barcelona' model had consolidated and become strongly characterized in terms of tourism growth (Arbaci and Tapada-Berteli, 2012; Degen and García, 2012). Indeed, a new surge in critical research on Barcelona's urban development has taken place since 'overtourism' has become a research buzzword on places. It could actually be claimed that the early works that introduced this term (Milano, 2018; Goodwin, 2019) were directly referring to the case of Barcelona. In this and many other Mediterranean destinations, the concern for tourism excesses intersects critically with the social effects of the global financial crisis of 2008. The economic slump in the Catalan city left a long trail of job losses and impoverishment, and a new dependency on tourism as a 'recovery strategy', which ultimately affected the most vulnerable strata of the population who have been helpless in the face of the revalorization dynamics in the housing market. It has also been greatly aided by the progressive deregulation of housing uses and the worsened employment conditions after the 'anti-crisis' labour reforms (Russo and Scarnato, 2018).

At the onset of the 2010s, the city presented itself as a global destination that also draws a significant number of day visitors from neighbouring regions – some of Spain's most popular resort areas – and it is an established haven for 'mobile populations' such as foreign students, displaced workers or transient migrants. For the first time in 2016, the city administration estimated that 28.1 million visitors was a realistic estimation of the size of the visitor economy, but also of the pressure generated by this population (Municipality of Barcelona, 2017). The growth of accommodation supply – captured by rates of bed places in all types of establishments per 1000 residents – intensified dramatically in the 'core' tourist areas over the last decade. The ratio of 163 beds per 1000 inhabitants registered in the Old City district in 2010 rose to 223 by 2015, and to 265 by 2019. At the neighbourhood level, Cócola-Gant and López-Gay (2020) calculated that in the Gòtic (Gothic) Quarter this ratio goes up to more than 600. This pressure has risen considerably also in other districts that in 2010 were still rather untouched by tourism.

In 2015, the supply of bed places in hotels was 67,603. However, this was now paralleled by an active supply of 59,614 bed places in STRs distributed over 16,700 dwellings, a staggering growth from marginal numbers since the appearance of intermediation platforms like Airbnb in 2012. While the growth of hotel supply is strictly regulated, the enforcement of regulations and the practical possibility to control the expansion of STRs has not kept pace (Lambea Llop, 2017), and Airbnb came to be tagged as the 'first accommodation chain' operating in the city in terms of supply. By May 2019, the enforcement of regulations and controls produced a slight contraction of this offer, but the steep growth of dwellings on offer as individual rooms in shared apartments, an unregulated modality of lodging until 2020 (A. Quaglieri-Domínguez *et al.*, 2021, unpublished data), almost balanced this decrease.

The impacts of this kind of growth on the local community need further consideration. Data on household income and rent distribution of resident communities reveal that the situation of such areas with a high density of STRs, mid-to-low-income areas in origin, has not deteriorated significantly under severe tourism stress, and has actually improved in the last few years. However, these areas are also among those having experienced the highest rates of population mobility and substitution. Thus, the image we are seeing regards a neighbourhood population that has substantially changed in the process (see Fig. 1.1a). Registered residents have declined in some of the most intensively tourist areas of the city, like the historical core of the Gòtic, where population loss has been about -14% between 2010 and 2014 (with some census tracts reaching -40%), while in others (like the multi-ethnic El Raval) or the middle-class Dreta de l'Eixample the trend has been reversed and they have actually gained population. By contrast, the low-income national and foreign-born, less-educated population has been pushed out. The number of residents started increasing again in 2015–2019 (Fig. 1.1b), yet the population flowing in to the most intensively 'touristified' areas of the city is quite different and could rather be characterized as a wealthier and highly educated new transnational middle class.

The dynamics of the real estate market have indeed been peculiar in the last decade. Purchase prices touched bottom in 2013 because of the fall in demand in the wake of the financial crisis, but economic recovery pushed prices up to pre-crisis levels around 2018. Rental prices reached a new peak in the same year and have continued growing in most neighbourhoods well into 2019 (notably not in the Old City), especially in areas that saw the most intensive development of Airbnb-mediated STRs (A. Quaglieri-Domínguez *et al.*, 2021, unpublished data). Cócola-Gant and López-Gay (2020) comment that the 'airbnbization' of housing since 2012 is just one dimension of the speculative pressures leading to neighbourhood change and social exclusion and is enmeshed in wider trends of financialization of real estate capital and commercial assets as the new frontier of neoliberal accumulation strategies. However, the spatial association of the rise of temporary dwellings, housing price inflation, and socio-demographic change in Barcelona and in its tourist core specifically, is remarkable. García-López *et al.* (2020) provide consistent evidence of the effect of Airbnb's expanding offer on real estate prices in Barcelona: at city level, the growth of STRs has a 1.9% effect on rents and 5.3% on transaction prices; yet in areas with a high intensity of Airbnb offers, the effect is 7% on rents and a remarkable 19% on transaction prices. García-López *et al.* also noted that after 2016 – when

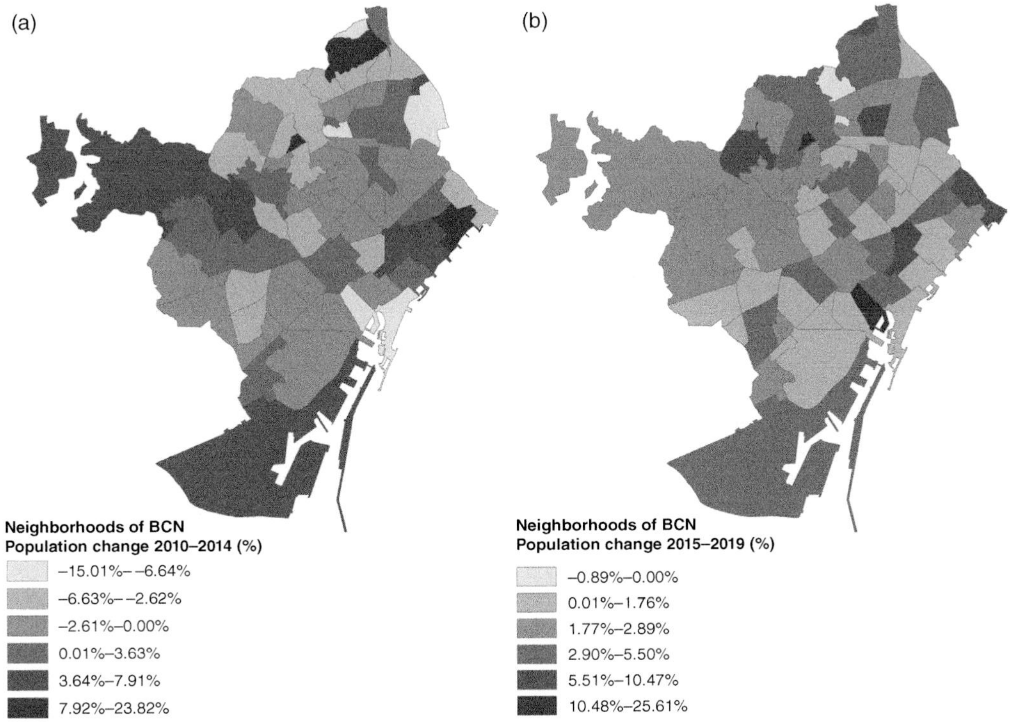

Fig. 1.1. Population change in Barcelona (BCN) neighbourhoods, 2010–2014 (a) and 2015–2019 (b). Adapted from Open Data Barcelona (Municipality of Barcelona, 2020).

new regulations and more pervasive controls started to curb the expansion of STRs – rental prices started to subside, yet purchase prices continued to grow, suggesting that speculative pressures continue to operate in the long term irrespective of such regulations.

Hence, a process of housing exclusion has been nuanced, which has at least in part been driven by the intensification of tourism. In 2017, the average effort rate to access housing for persons younger than 25 years old (share of personal income needed to afford a house) was above 125%: in other words, the average Barcelonian in this age cohort cannot afford to live in the city independently or needs to use other sources such as family savings or sharing the house with other adults. The rent prices obtained by landlords who let their property in the short-term accommodation market are between 2.3 and 4.1 times higher than those for long-term residents. Several authors note that this fact, together with the low supply of subsidized housing in Barcelona (less than 2% of the housing stock), is a key driver of socio-spatial segregation. However, the displacement of low-income population has been less apparent as homeownership has been promoted more than renting in Spain. The situation is different in regard to people who live in rental housing. Socio-spatial polarization in Barcelona is nothing new, as the historical city centre has been characterized by concentrations of poverty. In recent years, though, indicators show that this phenomenon seems to be receding, but this may well be due to the fact that middle classes have also been pushed to seek out neighbourhoods with a lower cost of living. In a 2012 survey, one in every six inhabitants of the metropolitan area of Barcelona reported to be suffering or having suffered housing exclusion of some sort (Nel·lo, 2018).

Vila de Gràcia: the enmeshments of tourism with social change

Vila de Gràcia may well be an exemplary case of the heterogeneity and complexity of the processes through which tourism has become a driver of social change. This neighbourhood, with 50,946 inhabitants in 2019, is one of five neighbourhoods comprising the district of Gràcia, which became officially incorporated into the City of Barcelona in 1897. Vila de Gràcia (henceforth: Gràcia) is the district's historical core. It is rooted in an historical identity of social and political activism, and a haven for transnational populations and visitors attracted by its cultural vibrancy and 'authenticity' (according to the results of a survey carried out by Barcelona City Council; Municipality of Barcelona, 2020), as opposed to the fading social identities of other historical areas of Barcelona. The choice of this neighbourhood for our study is based on the premise that, as opposed to other central areas of Barcelona, Gràcia is not a typical mass-tourist landscape, like the Gòtic Quarter, where economic and class differences between host and guest populations are vast. The area also features a relatively higher share of Airbnb supply in relation to other forms of commercial accommodation, a consequence of the fact that its tight urban fabric and planning regulations have impeded large-scale hotel developments. Gràcia, unlike the Gòtic, remains a densely populated neighbourhood: the 'touristification of housing' and the progressive reconfiguration of residential spaces through temporary dwelling practices can be seen to represent a hindrance only for specific groups of the incumbent population, enmeshed with a remarkable process of population substitution, or gentrification in the traditional sense, that took place in the last 10–15 years.

In what follows we carry out a comparative illustration of population trends, gentrification pressures and social perceptions on tourism development in Gràcia and the Gòtic against the global situation of Barcelona. We use various secondary sources reflecting a chronological development from 2016 to 2019. These include socio-demographic and migration data from the Open Data Service and from the Department of Statistics and Data Dissemination of the Barcelona City Council, as well as data on STRs that were obtained from providers like Inside Airbnb (2020).

This comparison fits our hypothesis of residential instability as a vector of social exclusion: both Gràcia and Gòtic are recognized tourist areas, with a high offer of accommodation, cultural facilities and a commercial landscape that suggest that a good part of their economy hinges on the expenditure of tourists. Yet the Gòtic is the paradigm of a core mass-tourist area, associated with mainstream practices (short visits, large groups, highly scheduled itineraries and related concentrations, focus on 'attractions', and use of services at the cheaper end), largely dictated by its central position and easy access from the beach and the port. It has been historically the area of concentration of mid-budget hotels and restaurants and has progressively been constructed as a mass-tourist consumption and transit area. By contrast, Gràcia is a relatively recent newcomer in the tourist map of Barcelona, mostly a result of its appeal for new visitor segments which are tuned to 'glocal' cultural landscapes and opportunities for social relations. Its location at the edge of the city centre proper, the lack of clearly defined 'visitor attractions', and the absence of a large hotel supply in the area have motivated an 'odd' pace of development and configuration as a tourist place in the last couple of decades.

This is reflected by the perception of local residents. On the one hand, in the 2019 municipal survey about their perceptions of tourism, 96.4% of respondents in Gòtic answered 'yes' to the question: 'Do you consider that you live in a tourist area?' By contrast, only 59.2% of respondents in Gràcia answered in the affirmative. On the other hand, the penetration of STRs in Gràcia is comparable to that of the Gòtic and so is the transformation of its commercial supply. In Gràcia, these trends are arguably tied to more traditional forms of transnational gentrification and population substitution where sheer 'touristification' plays a more limited and nuanced role.

Both neighbourhoods experience among the city's highest rates of residential mobility, or in other words, of instability in housing tenure. However, we can pick significant differences in their profile as 'attraction areas'. Figure 1.2 compares the migratory dynamics in the Gòtic, the mass-tourist heart of Barcelona, and in Vila de Gràcia, together with the total for Barcelona. The former presents a higher rate of population mobility and reflects much higher dynamics of immigration in these last few years with respect to Vila de Gràcia.

The flow of emigrants from Vila de Gràcia is almost constantly led by Spaniards, reflecting the general trend in the whole of Barcelona, while in the Gòtic there are larger emigration

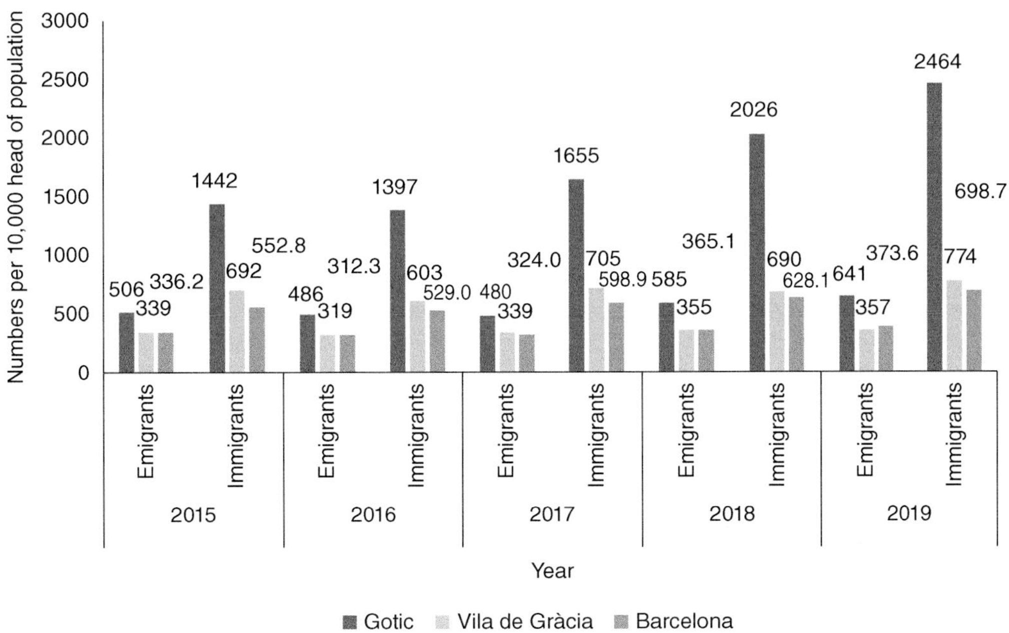

Fig. 1.2. Immigrants and emigrants 2015–2019 in (a) Gòtic neighbourhood; (b) Vila de Gràcia neighbourhood; (c) Barcelona total. Adapted from Barcelona City Council data (Municipality of Barcelona, 2020).

rates of foreigners. This could hint at a more traditional process of gentrification taking place in Gràcia, which is enmeshed in a subtler way with the recent explosion of STRs than in the Gòtic, where the substitution of dwellings occupied by local permanent residents with those occupied by tourists is much more pronounced. Comparing the demographics of these two tourist areas, the Gòtic is globally experiencing a rise in the relative share of the 25–34-year-old age cohort, while Gràcia has maintained a more balanced mix of younger and older aged individuals in the population.

In addition to age, the education levels of the resident population have been given emphasis in the gentrification debate as an indicator of social change. A comparison between the attainment levels in the two neighbourhoods also highlights opposite trends. The share of university graduates increased in Gràcia from 2015 to 2019, while in the Gòtic this segment decreased and the segment of the population with primary education as the maximum attainment level grew. This is another sign of 'traditional' gentrification taking place in Gràcia: the share of professionals and graduate students is rising in the area, intersecting with the dynamics of nationality. Most of the foreign immigrants from the neighbourhood have a higher qualification: two-thirds of these immigrants have university studies, as does the Spanish population that comes from outside Catalonia, while only half of the Spanish population within the neighbourhood has university degrees.

Both neighbourhoods have been heavily interested by the growth of the offer of STRs in the last decade. However, the Gòtic is also the area of greatest concentration of traditional hotels and other forms of 'regulated' hospitality, supporting the evidence of Gutiérrez *et al.* (2017) that the geography of Airbnb (and specifically that of whole apartments in STRs) tends to replicate that of the mainstream supply around the main attractions. In Vila de Gràcia (unlike other neighbourhoods of the district of Gràcia) the supply of hotels is not at all abundant, and the growth in popularity of the neighbourhood coincides with the period of expansion of the offer of STRs mediated by platforms. It is also, as seen above, an area hosting a sizable share of foreign middle-class residents who in previous research (e.g. A. Quaglieri-Domínguez *et al.*, 2021, unpublished data) are considered more likely to engage in 'shared hospitality' practices.

Thus, contrary to the rest of the city of Barcelona, where 'traditional' forms of accommodation are still dominant despite the large increase of STRs, in the case of Gràcia by far the largest share of tourist supply is represented by apartments on STR (corresponding to more than 95% of establishments and almost 60% of bed places). The rentals available on the Airbnb platform in May 2019 reached a total of 1583 dwellings in the Gràcia district, 1018 of which are in the Vila de Gràcia neighbourhood, including apartments fully or occasionally used as tourist accommodation, multiple rooms within the same apartment, and shared rooms (54.1%, 44.1% and 1.8%, respectively). In Fig. 1.3 the density of STRs of different types on offer in Gràcia is compared to that of the Gòtic. The Gòtic presents a denser distribution: there are 1184 dwellings on offer in Airbnb, against 1018 in Vila de Gràcia, which is more populated. This implies that in Gràcia there are 20 Airbnb dwellings for every 1000 inhabitants, while in the Gòtic this proportion rises to 62 per 1000 inhabitants, a pressure which is approximately three times larger.

While in Gràcia 54.1% of STRs are entire apartments, this figure only reaches 43.2% for the Gòtic. This can be interpreted as a reorientation of the market towards low-budget visitors and students who are more likely to use the abundant offer of individual rooms in STRs concentrated in the Gòtic. There is also evidence for this fact in that the average price of the dwellings in Airbnb is €84 in the Gòtic against €153 in Gràcia: on average, it is twice as expensive to stay in a tourist rental in Gràcia. This reaffirms the idea that Gòtic represents a traditional tourist core, while Vila de Gràcia is an emerging location in the market. It also shows that Gràcia, as opposed to the Gòtic, is being constructed as an area attracting a new type of tourist, who wants to experience authenticity and an everyday landscape, albeit off the beaten (tourist) track, avoiding 'tourist traps' and spaces that have been designed almost exclusively for visitor consumption.

The Impact of the Increasing Penetration of Tourism Dwellings in Gràcia

Methodology

In this section we present the main results of our investigation on the profiles, perceptions and motivations of current and former residents of Vila de Gràcia in the face of the tourism-related transformations of the neighbourhood and in particular residential and commercial change. In this section we have used secondary sources – the

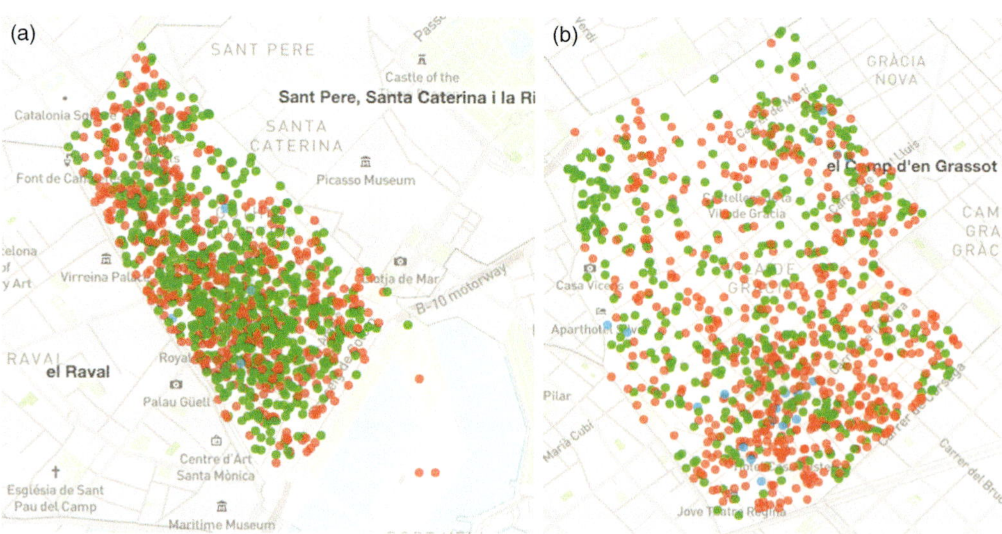

Fig. 1.3. Rentals available on the Airbnb platform by typology for Gòtic (a) and Vila de Gràcia (b). Whole apartments are in red and rooms for rent are in green. Adapted from data from Inside Airbnb (2020).

municipal residents' survey of perceptions on tourism (years 2016–2019) – and information obtained from our own field interviews. As for the former, the microdata have been elaborated to focus on specific profiles of respondents in different years in Gràcia. When it comes to primary sources, the data emerge from a survey conducted between the end of June and the end of July 2020. The survey included respondents among current and former residents of Vila de Gràcia, as well as members of neighbourhood associations and grassroots entities. In total, 29 Spaniards and 17 foreigners were surveyed. They were aged between 20 and 60 years, of different gender, age and education level, with quotas of age and education in correspondence with population distribution. The survey which we conducted contained 16 questions of opinion and perception, together with other profiling information (Hernández Sampieri, 2010). The completion of the questionnaire took around 30 minutes. The sample with which we worked was random and not probabilistic. The size of the sample was delimited according to the objectives pursued in the research (Hernández Sampieri, 2010). Of these 46 completed questionnaires, 15 led to in-depth open interviews, which allowed us to obtain further qualitative information, including again current neighbourhood residents and members of the neighbourhood council. In total these interviewees were 12 Spaniards and three foreigners, aged between 23 and 47 years. We asked them about issues such as perceptions and forms of coping with mass tourism, housing for tourist use and the effects on neighbourhood and personal lives.

Community perception of tourism

The Barcelona City Council has been carrying out regular yearly surveys of public opinion on tourism, gauging the residents' perception on issues such as the benefits produced by tourism, the public management of tourism, hindrances to quality of life and access to housing in the context of tourist city, among others. The results of these surveys prove the rising controversial character of tourism (Russo and Scarnato, 2018) among the Barcelonians, who for the first time in 2017 perceived tourism as the biggest source of problems for their quality of life. It is, however, interesting to contrast the opinions of residents in the most tourist-trodden areas of the city and, in this case, in the Gràcia district. We have characterized the latter area as moderately affected by mass tourism and one in which a tourist seems to blend in more smoothly in the sociocultural fabric of the place rather than what is observed in the mass-tourist core of the city. Thus, the responses to the resident survey in the 2016–2019 editions (Fig. 1.4) show a declining rate of approval for tourism in Gràcia, with a remarkable slump between 2018 and 2019. However, there is still a majority of the population who, evaluating benefits and problems generated by tourism, are ready to express an overall positive opinion. Arguably, the softer or the more indirect the impacts of tourism are perceived in the neighbourhood of residence, the greater the approval appears to be.

Rising dissatisfaction may well be related to the unbridled growth in tourism activity and its effects on neighbourhood change and the process of commodities and housing prices. This was revealed through specific questions regarding the perception as to whether tourism growth is responsible for price inflation in the city and in the neighbourhood of residence of the respondent. While residents of a tourist-ridden neighbourhood are prone to notice the inflationary effect of tourism on the city as a whole, they are much likelier to note this effect in their own neighbourhood, with an ample difference with the city average. In Gràcia the perception of an inflationary effect of tourism on their neighbourhood is around 23 percentage points higher than the perceived inflationary effect of tourism on the city as a whole. However, residents of Gràcia are less critical about the effects of tourism on their neighbourhood's cost trends than those of the Gòtic, while they are more likely than those in the Gòtic to affirm that tourism has an effect on prices of the city as a whole; this opinion has been growing in predominance among Gràcia's residents over the period 2016–2019. This could be interpreted as a late reaction of Gràcia's residents on the perceptions of the intensification of tourism in the city and in their neighbourhood, compared to those in the Gòtic, who have been exposed to this process for a longer time. Yet it might also reveal the fundamental difference in the profile of respondents, a

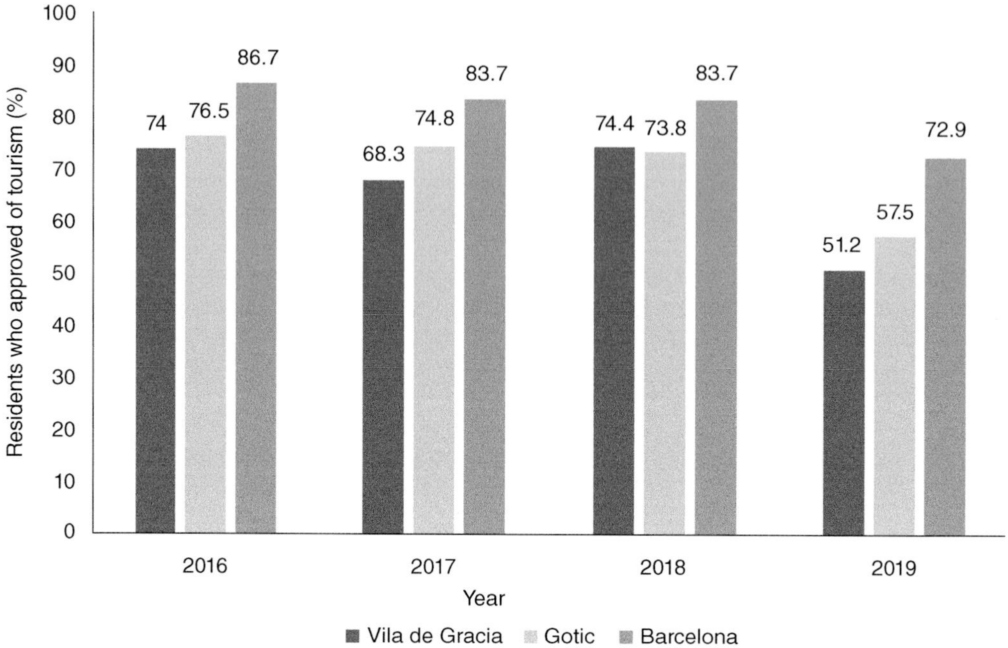

Fig. 1.4. Level of approval of tourism in the city by residents of Vila de Gràcia, Gòtic and Barcelona city. Adapted from Open Data Barcelona (Municipality of Barcelona, 2020).

consequence of the processes of population change described earlier, by which the residents who are directly affected by price inflation are relatively fewer.

The flourishing of an offer of retail stores, restaurants, cafés and boutiques in Gràcia, clearly targeting cosmopolitan consumers, is a clear further indication of classical gentrification. The new commercial supply is, in itself, an agent of social change; in spite of its good quality, it takes away from the social and consumption habits of the incumbent population whose purchasing power limits what they can afford. However, in a context of high population mobility as illustrated above, this new supply serves affluent incomers as well as visitors. Another of the municipal survey questions gauged the perception of residents concerning the rising profile of the visitor economy in their neighbourhood. The score relating to this issue is stable among respondents from Gràcia over the control period and doubles approximatively the figure of Barcelona as a whole. However, it is around 27 points less than with the respondents from the Gòtic, though this gap seems to be shrinking, perhaps because of the ongoing implementation of regulations which has curtailed the steep growth in the Gòtic. This and other policies and regulations implemented by the municipal and Catalan government have been focusing on ensuring benefits for the tourist and retail sector while maintaining a high quality of life for native residents. Yet, residents continue to express dissatisfaction with the effectiveness of these programmes. Responses to one of the survey questions, the approval of public management of tourism, highlighted that in spite of the moderate positive response to the regulations introduced by the current administration in its first mandate (from a negative balance of -1.9 to a positive one of +5 between 2016 and 2017), in 2019 there is a clear turn of public opinion in this respect, dropping to -15.2 from +5.1 the previous year.

Thus, the public perception of tourism in the Vila de Gràcia neighbourhood presents different faces. Although there is value in the regeneration and brand attractiveness of the area, and in the economic activity that this brings, signs of dissatisfaction with the impacts of tourism – which need filtering with a focus on population change – seem to be on the rise.

Results from the fieldwork

As seen above, the population in Gràcia is subject to dynamics that represent traditional gentrification. At the same time, these inhabitants resent tourism 'intensification' as a key driver of neighbourhood change, entangled with socio-economic changes linked to the promotion of the place as a tourist attraction. The sample of participants in our field interviews replicates these dynamics: only 26% of the respondents have lived more than 5 years in Gràcia, while 6.5% have been living there for less than 1 year. The largest group is composed of residents who have lived there from 3 to 4 years.

A respondent, a woman with 8 years of residence in the neighbourhood, commented: 'I know cases of families who have lived more than 20 years in the neighbourhood and have left. At present, many young people who arrive in the neighbourhood stay no longer than 3 years'. This impression ticks with the evidence that most residents limit their stay in the neighbourhood to a couple of years, while others who have maintained their domicile for many years, under the current situation, prefer to settle in another district that offers them greater opportunities and quality of life according to their economic possibilities. By contrast, regarding displacements related to transnational migrations, a 43-year-old, who had lived 5 years in the Vila de Gràcia neighbourhood, commented that:

> The Vila de Gràcia neighbourhood is pretty multicultural, there are mostly Spaniards, but it is also clear that there are many foreigners of all nationalities ... In my residence you can no longer relate to the neighbours because many of them are foreigners and tend to team up with their own people.
>
> (Interviewee 1)

Among survey participants who previously lived in Gràcia and decided to move to another district, the majority reported the rising rents as the main driver of their decision. Yet almost 37% of them recognize that the excessive presence of tourists has to do with relocation decisions. We have seen above how decisions to relocate out of Gràcia affects primarily the population of Spanish origin.

Interviewees with a longer residence time in the neighbourhood pointed out that urban renewal in the area is mainly focused on the attraction of visitors. Along the same lines, residents with a shorter residence track in Gràcia acknowledge that there have been a few large-scale renovations, and yet these all seem to take place for the benefit of tourism. A 35-year-old Spanish interviewee expressed:

> The neighbourhood has improved a lot in recent years both in mobility and in building rehabilitation. Tourism has had a lot to do with it, as many owners have chosen to reform homes to use them for this purpose and to obtain better rental yields.
>
> (Interviewee 2)

The renewal of the area is seen to take away the symbolic landscape that contributes to the identity of the area, encouraging different circulations and flows of inhabitants. This restructured landscape also offers an image of modernization and progress that attracts new enterprises, activities and land uses to the neighbourhood, producing fertile economic dynamics, revitalizing many aspects that stayed unchanged for a long period. The new businesses and establishments in Vila de Gràcia are aimed at a transnational tourist and migrant population but also have a positive impact on locals who have access to a greater array of goods and services. In our interviews, the positive and negative perceptions in this regard balance out. Most interviewees acknowledge that new businesses are focused on tourism. A 33-year-old Spanish, 5-year resident of the neighbourhood, explained:

> All the old premises have been converted into bars and restaurants or international brand outlets. For example, they have menus in multiple languages, and the main requirement for hiring staff is proficiency with foreign languages, rather than experience. On several occasions I have to go to other areas to be able to buy day-to-day products.
>
> (Interviewee 3)

According to our survey, the great majority of participants (more than 78%) state that the distribution of the businesses is equitable, since they cover the needs of the local resident and the tourist, while a minority claim that businesses are aimed at attracting visitors with the greatest number of businesses oriented at the tourism sector.

These results illustrate the ambiguous effects of tourism growth in an area. On the one hand, it pushes up housing prices and expels population. On the other hand, it creates new opportunities for business and the appearance of new needs and new standards of living for local residents. With time and the incorporation of new habits and practices, the standard of services required by tourists and local residents start to converge again. Yet, the displacement of many residents may have an unsettling effect on those who stay in the neighbourhood. The possibility of orienting real estate to tourist use with a higher rentability, affects not only prices but also renting conditions, as was made evident in some of our interviews. The 43-year-old Spanish resident with a 5-year residence in the neighbourhood stated:

> Obviously before we could rent an apartment for a family with a decent and secure contract that ran from 5 to 8 years with extensions. Nowadays everything has changed and the contracts they make to you now are garbage, they do not provide us with security for those of us who have families, the contracts they make go between 1 to 3 years and with very high prices. The reason seems to me is because the real estate or apartment owners realized that renting the apartment by the day or by rooms is more profitable per month.
>
> (Interviewee 1)

Other interviewees point their finger at the decrease in available long-term rentals and an increase in prices, as STRs become a more lucrative option for owners: 91.3% of the participants stated that the increase in STRs in the neighbourhood had an impact on consumer and rental prices, on the overcrowding of public spaces, noise pollution, the environment and insecurity.

In relation to the difference in the cost of long-term rentals compared to STRs, the interviewees expressed that this is a matter of the economic market, since long-term rental is not as profitable as STR. A 29-year-old Spanish resident with less than 1 year of residence in the neighbourhood, said: 'What a shame to profit from this poorly controlled market taking advantage of regulatory gaps, while other regulated hotel and extra-hotel establishments are paying the consequences' (Interviewee 4). In fact, the increase in this type of accommodation has not only harmed the residents through limiting accessibility to long-term rentals and the enhanced tourist boom in their neighbourhood. It has also had an evident impact on traditional accommodation since all commercial activities in the area have increased their prices.

As a complement to the interviews carried out, the participants were asked if, while they were living in the Vila de Gràcia neighbourhood, they received any offers to purchase the property from a private or public investor. Two of the participants said this had happened but they declined the offers. A 31-year-old Spanish person with 7 years of residence in the neighbourhood, expressed: 'I have received several offers for the purchase of my apartment from individuals or agencies that sell properties without specifically explaining the reason for their interest and insistence' (Interviewee 5). This provides evidence that the pressure exerted by investors seeking the acquisition of new properties may be a more general trend, in those areas where 'buying to let' in the STR market is a flourishing new investment opportunity.

Getting used to the new reality, coping with it by generating resilience, as well as refusal and rejection are typical responses to abrupt changes. According to our survey, 67.4% of the interviewees stated that the resident–tourist relationship in Gràcia is 'Regular, with little cordial gestures between the two groups'. This resonates with the results of another municipal survey-based study, the victimization survey of 2019, which reports that the highest percentage of conflicts attributed to tourism behaviour is reported by residents of the district of Ciutat Vella (57.6%), followed by Gràcia (37.2%). A 45-year-old Spanish woman who has been a resident of the neighbourhood for 7 years, explained:

> The relationship with tourists is to ignore each other. Personally, when I see so many tourists in the neighbourhood, I am overwhelmed and I do not behave very cordial, since due to their arrival I will not be able to sleep because of the noise, find the streets dirty, walk carefully through the streets because they walk drunk, some do their needs [urinate] at the doors of the building or even some inside, also in the long term, influencing the rise in rental prices and surcharges throughout the neighbourhood.
>
> (Interviewee 6)

The rest of the interviewees pointed out that for the most part, relationships are good as long as tourists behave. Yet Gràcia residents can get overwhelmed with the excess of tourists. They suffer from the noise produced by the night-time visitor economy, but also from public disorder, littering and disrespectful behaviour, and this increasingly interferes with their willingness to have contact with tourists. Former and current residents of the neighbourhood point out the personal changes they have suffered in recent years, highlighting the increase in stress due to overcrowding and noise pollution due to the lack of night rest and the increasing instability of their position as tenants. Through the survey carried out with former residents and members of neighbourhood associations, it was found that 26% characterize their experience of residing in Gràcia in terms of poor quality of life, 34.8% with 'regular' and 39.2% with 'good' quality of life. The 43-year-old Spanish woman who has lived in the neighbourhood for 5 years, states to this respect:

> The standard of living in the neighbourhood has declined because its true roots have been lost, they have changed their true identity to make way for a new era of tourists; thanks to whom everything became more expensive and the people who have been living here for years, including some retirees, only collect their small pension, which is not enough to live with dignity in their own neighbourhood because life became more expensive overnight.
>
> (Interviewee 1)

The idea of the loss in the true identity of the neighbourhood is also present in other statements of respondents. A young clerk of 23 states: 'It is sad, but the tourists have turned the area into what it is today. When they crawl around their streets, the essence and identity of the neighbourhood is lost, step by step' (Interviewee 7). The point is that tourists experience spaces in a different way from that of residents. As it can be interpreted from these testimonies, tourists deprive locations of their identity: touristification thus implies a sort of uniformization of the landscape and the erosion of local idiosyncrasy that has been socially constructed by a 'stable' population. As many as 63% of our sample of informants agree with the statement that 'it is possible the neighbourhood may suffer from tourist overload' and have observed a rising predominance of a transitory foreign population in comparison to Spanish 'stable' residents. The Spaniard with 7 years of residence in the neighbourhood, stated:

> Everything is out of control now, and the tranquillity that previously reigned in the neighbourhood has been completely lost; now we suffer from several disruptions and accidents related with the presence of tourists in the neighbourhood and the tourist use of homes in our buildings.
>
> (Interviewee 5)

Daily activities such as going to the supermarket or sitting at your favourite bar are increasingly hindered by crowds, and rest is disrupted by noisy and loud passers-by. Among the pressures experienced by residents, the rise in housing prices is the main factor in influencing decisions to move away from the neighbourhood. A 44-year-old Spanish interviewee with a 5-year-old residence in the neighbourhood stated:

> There should be a limit; there is a balance in the rent in the short and long term. Residents have been forced or [are] in the need to leave our neighbourhoods due to the increase in rental prices, while the owners get rich with tourist rentals; where residents significantly lower their quality of life in large cities, having to move to rural areas.
>
> (Interviewee 8)

Along this line, 95.7% of those surveyed in the Vila de Gràcia neighbourhood stated that government management to regulate homes for tourist use is insufficient or not effectively implemented. About government management, the 31-year-old Spanish resident who has lived in the neighbourhood for 7 years, said:

> If what the Barcelona City Council really wants is to control illegal flats, they are not doing a good job. They have decided to cut all tourist accommodation (including luxury hotels) in the area, instead of focusing on having real control over tourist apartments.
>
> (Interviewee 5)

Most of the interviewees affirm that corrective measures should be taken to solve the problems that the intensification of tourism activity in the neighbourhood has imposed on the quality of life of the residents. The 43-year-old Spanish resident for 5 years, said: 'The City Council and the Government of Catalunya should pay more attention and therefore work on clear laws, which currently do not exist, in order to better control the phenomenon of tourism in general' (Interviewee 1). Some respondents point to the lack of political will to control this situation and reverse negative trends. A Spanish female tourism professional stated: 'It is a pity that we reached this scenario because of the lack of anticipation and solutions. I imagine this is the price to be paid for being global and European leaders of tourism, at least before the pandemic' (Interviewee 9). Here we come across the idea of 'a price to be paid', that is, the rise in the prices of goods and services and the appearance of speculative pressures on the real estate market of the neighbourhood as the inevitable side effect of investments and jobs created in tourism and hospitality, and the related commercial dynamics. Another resident claims that tourism is overvalued as the silver bullet to overcome economic crises of the past and the future. According to this respondent, a 35-year-old female hotel receptionist:

> It is a topic that we have already sensed since the moment the first tourists stepped foot here, and all the businessmen saw they had the hen of the golden eggs. Those who paid for the broken glasses were us, the residents, who are obliged to pay for stuff as if we were also tourists.
> (Interviewee 10)

Discussion and Conclusions

Our study has related the changing social configuration of Vila de Gràcia, its residents' perceptions of tourism, and the personal life stories on how our interviewees have been affected, resituated or coped with the tourist transformation of their neighbourhood. In particular, we have examined one of the main drivers behind this transformation, the dramatic rise of STRs on offer that took place in recent years. This storyline allows nuancing social change in a neighbourhood that is possibly not 'overtouristed' in the strict semantic connotation given by the recent surge in research. Contrary to other 'core' tourism areas of Barcelona and many other historical centres of global destinations, Gràcia has maintained itself at the margins of the typical phenomena related to excess tourism pressure – overcrowding, littering, the impoverishment and banalization of the commercial offer, and ultimately the displacement of a sizable share of its residents. Due to its relatively peripheral situation and to the inconvenience of its physical fabric for large-scale hotel development, Gràcia has been able to maintain its identity traits and a strong sense of in-bred cultural activism and neighbour solidarity. Paradoxically, however, these very features have put it on the map of attractive places for a new generation of 'mobile dwellers' who, on the one hand, veer off the beaten track and, on the other, seek to 'plug in' to a vibrant cultural landscape. These individuals include new foreign residents, like highly skilled expats, members of the academic community and other types of global nomads. Yet, tourists have also been able to profit from the new opportunities opened by the boom of platform hospitality to reside short term in the neighbourhood. They are lured by Gràcia's newly gained prominence as a 'place to be' and by the effect of the rising residential and commercial gentrification. The two storylines interlink with each other since, as has often been noted, real estate investments by transnational gentrifiers go hand in hand with 'buy-to-let' practices and hospitality platforms, which have become the main agents in the production of new area images that appeal to the international travel market.

Thus, tourism has inexorably crept into Gràcia in spite of its characteristics. It is a very different type of tourism from that of the tacky central neighbourhoods like the Gòtic, which we have used in comparison. Its social fabric is quite different. It includes a middle class already sizably intermingled with the recently gained transnational population. While the sheer numbers reveal an increasing process of population substitution and internationalization rather than displacement and shrinkage, the effects on the incumbent local population are none the less problematic. The basic services for daily consumption, a pillar of relational social reproduction (Crespi-Vallbona and Mascarilla-Miró, 2018), have progressively given way to a 'glocalized' landscape, which is attuned to the needs of the recent wave of dwellers. In our interviews, this seems to be the most important factor that may ultimately drive traditional residents out of the area, together with the rising price of rents and purchase prices.

Yet commercial change is not the only avenue that may push a part of the local community at the margins of social reproduction. Other subtler effects are taking place, as revealed by our small sample of interviewees. Remarkably, all these processes revolve around the 'touristification of housing' and the emergence of transient forms of 'dwelling' (rather than residing in a stable way) in an area. These include: (i) casual workers increasingly employed as keyholders and service providers to STRs, but also made invisible by a sector that uses the rhetoric of 'authentic local hosts' as differential value; (ii) working-class families or elderly persons who live alongside visitors staying in home spaces and leisurely moving around these homes; or (iii) tenants whose life stability and personal and professional choices are challenged by the knowledge that their home could easily become the next investment opportunity for their landlords. Even before they are actually displaced by the housing market dynamics, these persons have voiced their discomfort concerning the ways in which the touristification of housing and of their everyday spaces have affected them: (i) the erosion of their local 'support systems'; (ii) the disruption of daily routines; and (iii) the increasing hindrances for their rest, security and safety. They feel out of place in the changing neighbourhood although they mostly acknowledge its newly gained economic vitality. At the same time, they seem to blame the municipal government for supporting these changes through public expenditure in the area renewal rather than doing enough to prevent the side effects they suffer in their personal life.

The case of Vila de Gràcia in our study is possibly one among many others, which provide insights into the reasons behind a rising confrontation between locals and visitors as they contest overdwelling and the consequences this contest may have in terms of social change when the 'losing' side – less capable to adapt to such reorientation of the neighbourhood – is ultimately displaced. As we have tried to show in this work, the autochthonous population of the neighbourhood, including the locals with long-established roots in the area as well as former generations of work-related migrants, are conscious of the rising hardships related to maintaining their residence in the area. Although remaining in the area may be their primary ambition, they also experience increasing difficulties in coping with the economic as well as material pressures determined by the rising transformation of their neighbourhood as a gentrified tourist place. Contrary to claims of 'tourismophobia', which are frequently used to deface the critical attitudes towards tourism in the public debate (Pirillo Ramos and Mundet, 2021), the participants of this study are not unhappy about living in a touristified neighbourhood and having to 'share space' with new dwellers, even when it directly affects their rest, consumption and mobility routines. However, they are also well aware that they are in an uncomfortable position given the possibility of being next in line to have to leave when their rented homes are converted into tourist apartments. The anxiety, stress and mood swings, which they associate with having to cope with an ever-shifting everyday life environment may also lead to decisions to abandon the area and seek out new residence in other neighbourhoods or cities where the tourist pressure is lower, and the attainability of everyday commodities can be adjusted to their needs.

These trends in Vila de Gràcia or elsewhere inform the debate about the resilience of places in the face of tourism mobilities, which could eventually contribute to a thorough reset of tourism development and planning policies (cf. Salerno and Russo, 2020). Mainstream approaches to resilience in tourism (e.g. Lew, 2014) invite us to consider that the tourism-driven gentrification of Gràcia is not, in any way, reducing its attractiveness as a destination and, therefore, a generator of prosperity to the local community: its 'essence' is maintained in spite of the loss of a stable autochthonous population, because that essence, anyway, is a contingent and constructed notion (Richards and Russo, 2016). However, a normative notion of resilience, based on the capacity of response of cities and communities in relation to the disruptive force of tourism on the citizens' well-being, should inform policy changes towards the legitimation of claims of 'right to the city' and to stable residence (Grove, 2014). In the context of the current pandemic crisis, which in Barcelona as elsewhere has unearthed the vulnerability of the city and its social and economic structures to the dependency from tourism revenues, building on this second 'subversive' notion of resilience may well represent the most urgent challenge to feed a process of recovery. The transitionary nature of tourism and the upscale consumption practices of new generations of transnational gentrifiers form the basis of social change in Gràcia. These types of place users are highly enmeshed within property regimes that increasingly tend towards short-term dwelling patterns (López-Gay *et al.*, 2020) and motivate a porous neighbourhood commercial supply. The blunt tools provided by municipal planning schemes cannot solve the progressive shift of commerce beyond the reach of what long-term residents can afford on a daily basis. However, most importantly, a basic dimension of such a drift towards the 'touristification of the everyday' – housing – has not been tackled for some time in its most problematic features either, and this is possibly the most radical agent of erosion of stable residence.

References

Arbaci, S. and Tapada-Berteli, T. (2012) Social inequality and urban regeneration in Barcelona city centre: reconsidering success. *European Urban and Regional Studies* 19(3), 287–311.

Ardura Urquiaga, A., Lorente-Riverola, I. and Ruiz Sanchez, J. (2020) Platform-mediated short-term rentals and gentrification in Madrid. *Urban Studies* 57(1). DOI: 10.1177/0042098020918154

Bialski, P. (2016) Authority and authorship: uncovering the socio-technical regimes of peer-to-peer tourism. In: Russo, A.P. and Richards, G. (eds) *Re-Inventing the Local in Tourism*. Channel View, Bristol, UK, pp. 35–49.

Cócola Gant, A. (2016) *Apartamentos turísticos, hoteles y desplazamiento de población.* Barcelona, CreativeCommons. Available at: http://orca.cf.ac.uk/96338/1/Informe.pdf (accessed 20 March 2021).

Cócola-Gant, A. and Gago, A. (2021) Airbnb, buy-to-let investment and tourism-driven displacement: a case study in Lisbon. *Environment and Planning A: Economy and Space* 53(7), 1671–1688.

Cócola-Gant, A. and López-Gay, A. (2020) Transnational gentrification, tourism and the formation of 'foreign only' enclaves in Barcelona. *Urban Studies* 57(15), 3025–3043.

Crespi-Vallbona, M. and Mascarilla-Miró, Ó. (2018) La transformación y gentrificación turística del espacio urbano. El caso de la Barceloneta (Barcelona). *EURE* 44(133), 51–70.

Degen, M. and García, M. (2012) The transformation of the 'Barcelona model': an analysis of culture, urban regeneration and governance. *International Journal of Urban and Regional Research* 36(5), 1022–1038.

Doxey, G. (1975) A causation theory of visitor-resident irritants: methodology and research influence. In: *The Impact of Tourism: the Travel Research Association Sixth Annual Conference Proceedings,* San Diego, California, September 8–11, 1975. Travel Research Association, Salt Lake City, Utah, pp. 195–198.

Dredge, D., Gyimóthy, S., Birkbak, A., Elgaard Jensen, T. and Madsen, A. (2016) *The Impact of Regulatory Approaches Targeting Collaborative Economy in the Tourism Accommodation Sector: Barcelona, Berlin, Amsterdam and Paris. Impulse Paper N. 9.* Prepared for the European Commission DG GROWTH, Aalborg University, Copenhagen.

Füller, H. and Michel, B. (2014) 'Stop being a tourist!' New dynamics of urban tourism in Berlin-Kreuzberg. *International Journal of Urban and Regional Research* 38(4), 1304–1318.

Garcia-López, M.À., Jofre-Monseny, J., Martínez-Mazza, R. and Segú, M. (2020) Do short-term rental platforms affect housing markets? Evidence from Airbnb in Barcelona. *Journal of Urban Economics* 119. DOI: 10.1016/j.jue.2020.103278

González-Pérez, J.M. (2019) The dispute over tourist cities. Tourism gentrification in the historic Centre of Palma (Majorca, Spain). *Tourism Geographies* 22(1), 171–191. DOI: 10.1080/14616688.2019.1586986

Goodwin, H. (2019) Barcelona – crowding out the locals: a model for tourism management? In: Dodds, R. and Butler, R. (eds) *Overtourism: Issues, Realities and Solutions* (Vol. 1). De Gruyter, Berlin, pp. 125–138.

Grove, K. (2014) Agency, affect, and the immunological politics of disaster resilience. *Environment and Planning D: Society and Space* 32(2), 240–256.

Gutiérrez, J., García-Palomares, J.C., Romanillos, G. and Salas-Olmedo, M.H. (2017) The eruption of Airbnb in tourist cities: comparing spatial patterns of hotels and peer-to-peer accommodation in Barcelona. *Tourism Management* 62, 278–291.

Hernández Sampieri, R. (2010) *Metodología de la investigación*. McGraw-Hill, México.

Hiernaux, D. and González, C.I. (2014) Turismo y gentrificación: pistas teóricas sobre una articulación. *Revista de Geografía Norte Grande* 58, 55–70.

Inside Airbnb (2020) About Airbnb in Barcelona. Available at: http://insideairbnb.com/barcelona/?neighbourhood=&filterEntireHomes=false&filterHighlyAvailable=false&filterRecentReviews=false&filterMultiListings=false (accessed 20 January 2020).

Jover, J. and Díaz-Parra, I. (2019) Gentrification, transnational gentrification and touristification in Seville, Spain. *Urban Studies* 57(15), 3044–3059.
Lagonigro, R., Martori, J.C. and Apparicio, P. (2020) Understanding Airbnb spatial distribution in a southern European city: the case of Barcelona. *Applied Geography* 115: 102136. DOI: 10.1016/j.apgeog.2019.102136
Lambea Llop, N. (2017) A policy approach to the impact of tourist dwellings in condominiums and neighbourhoods in Barcelona. *Urban Research & Practice* 10(1), 120–129.
Lew, A.A. (2014) Scale, change and resilience in community tourism planning. *Tourism Geographies* 16(1), 14–22.
López-Gay, A., Cócola-Gant, A. and Russo, A.P. (2020) Urban tourism and population change: gentrification in the age of mobilities. *Population Space and Place* 27: e2380. DOI: 10.1002/psp.2380
Milano, C. (2018) Overtourism, malestar social y turismofobia. un debate controvertido. *Pasos* 16(3), 551–564. Available at: https://www.pasosonline.org/Publicados/16318/PS318_01.pdf (accessed 20 March 2021).
Milano, C., Cheer, J.M. and Novelli, M. (eds) (2019) *Overtourism: Excesses, Discontents and Measures in Travel and Tourism*. CAB International, Wallingford, UK.
Monclús, F.J. (2003) The Barcelona model: and an original formula? From 'reconstruction' to strategic urban projects (1979–2004). *Planning Perspectives* 18(4), 399–421.
Municipality of Barcelona (2017) *Estratègia de mobilitat turística de Barcelona*. Available at: https://ajuntament.barcelona.cat/turisme/sites/default/files/memoria_emt_20171204.pdf (accessed 7 November 2021).
Municipality of Barcelona (2020) Open Data BCN. Ajuntament de Barcelona's open data service. May 2020. Available at: https://opendata-ajuntament.barcelona.cat/en (accessed 10 November 2020).
Nel·lo, O. (2018) Segregació residencial, gentrificació urbana i dret a la ciutat: una lectura des de Barcelona. *Papers: Regio Metropolitana de Barcelona* 60, 49–65.
Novy, J. (2019) Urban tourism as a bone of contention: four explanatory hypotheses and a caveat. *International Journal of Tourism Cities* 5(1), 63–74.
Petruzzi, M.A., Marques, G.S., do Carmo, M. and Correia, A. (2020) Airbnb and neighbourhoods: an exploratory study. *International Journal of Tourism Cities* 6(1), 72–89.
Pirillo Ramos, S. and Mundet, L. (2021) Tourism-phobia in Barcelona: dismantling discursive strategies and power games in the construction of a sustainable tourist city. *Journal of Tourism and Cultural Change* 19(1), 113–131.
Richards, G. and Russo, A.P. (2016) Synthesis and conclusions. In: Russo, A.P. and Richards, G. (eds) *Reinventing the Local in Tourism*. Channel View, Clevedon, UK, pp. 251–266.
Richards, S., Brown, L. and Dilettuso, A. (2019) The Airbnb phenomenon: the resident's perspective. *International Journal of Tourism Cities* 6(1), 8–26.
Russo, A.P. and Scarnato, A. (2018) 'Barcelona in common': a new urban regime for the 21st-century tourist city? *Journal of Urban Affairs* 40(4), 455–474.
Salerno, G.M. and Russo, A.P. (2020) Venice as a short-term city. Between global trends and local lock-ins. *Journal of Sustainable Tourism.* DOI: 10.1080/09669582.2020.1860068
Segú, M. (2018) Do short-term rent platforms affect rents? Evidence from Airbnb in Barcelona. MPRA Paper No. 84369. Available at: https://mpra.ub.uni-muenchen.de/84369/1/MPRA_paper_84369.pdf (accessed 20 March 2021).
Semi, G. and Tonetta, M. (2020) Marginal hosts: short-term rental suppliers in Turin, Italy. *Environment and Planning A: Economy and Space* 53(7), 1630–1651. DOI: 10.1177/0308518X20912435
Sequera, J. and Nofre, J. (2018) Shaken, not stirred: new debates on touristification and the limits of gentrification. *City* 22(5–6), 843–855.
Smith, A. (2005) Conceptualizing city image change: the 're-imaging' of Barcelona. *Tourism Geographies* 7(4), 398–423.
Vale, L.J. (2014) The politics of resilient cities: whose resilience and whose city? *Building Research & Information* 42(2), 191–201.
Wachsmuth, D. and Weisler, A. (2018) Airbnb and the rent gap: gentrification through the sharing economy. *Environment and Planning A: Economy and Space* 50(6), 1147–1170.

2 Social Exchanges and Peer-to-peer Accommodation: Residents' Perceptions in a Neighbourhood Context

Dimitrios Stergiou,[1]* Anna Farmaki[2] and Fani Efthymiadou[2]
[1]Hellenic Open University, Greece; [2]Cyprus University of Technology, Cyprus

Abstract
The rapid growth of peer-to-peer (P2P) accommodation has challenged the sustainability efforts of destinations, due to the numerous negative impacts accrued from the influx of tourists in residential areas. Even so, scant attention has been paid to the perceptions of local residents with regard to associated impacts. This chapter aims to examine the perceptions of the residents of Koukaki (Athens, Greece) regarding the impacts of Airbnb at the neighbourhood level, using social exchange theory. Findings reveal a dominance of negative perceptions of socio-economic and environmental impacts and identify attempts from residents to mitigate associated impacts. Drawing from the literature and empirical findings, the chapter offers a framework which acknowledges the mediating role of platforms and property owners in the social exchanges between residents and guests. The chapter concludes with a discussion of implications, contributing towards the adoption of an anthropocentric and inclusive approach on P2P accommodation.

Keywords: Airbnb, P2P accommodation, neighbourhoods, impacts, social exchange theory

Highlights

- Drawing from social exchange theory, we examine residents' perceptions of the impacts of the growth of Airbnb on local communities.
- Findings reveal a dominance of negative perceptions with residents identifying a range of socio-economic and environmental impacts.
- The chapter discusses ways in which residents try to mitigate the platform's negative impacts on their local community.
- Findings also highlight the importance of the mediating role of the platforms and property owners in the social exchanges of residents and guests.

Introduction

Peer-to-peer (P2P) accommodation is growing phenomenally in scale and scope, changing the dynamics of the tourism and hospitality industry in both positive and negative terms. The spectacular growth of Airbnb, the market leader of

*dstergiou@eap.gr

(eds Anna Farmaki, Dimitri Ioannides and Stella Kladou)
DOI: 10.1079/9781789246605.0002

P2P-accommodation networks, exemplifies this phenomenal expansion and corresponding impacts accruing to destinations. Founded in 2008, today Airbnb offers more than 7 million listings in over 200 countries (Airbnb, 2020). In this growth context, Airbnb is credited with a number of positive impacts. For example, Airbnb can bring additional income to hosts, create positive employment effects on local economies, increase the range of accommodation options for travellers and offer a more authentic tourist experience (Tussyadiah and Pesonen, 2016; Guttentag and Smith, 2017; Dogru *et al.*, 2020). Notwithstanding this, the growth of the platform has raised concerns over the immoral behaviours of hosts, the illegal activity of many rentals as well as the adverse effects on the hotel sector and local housing markets (Gutiérrez *et al.*, 2017; Hajibaba and Dolnicar, 2017; Farmaki *et al.*, 2019). In particular, criticism emerged over the intensification of Airbnb's presence in residential areas (Ioannides *et al.*, 2019) which has been acknowledged as contributing to overtourism (Paulauskaite *et al.*, 2017; Stors and Baltes, 2018) while exerting an array of problems at the neighbourhood level.

Within this context, several observers have argued that the popularity of Airbnb challenges the sustainability effort of many destinations, contradicting the initial suggestions that the sharing economy model presents a favourable route towards sustainable development (Martin, 2016; Cheng *et al.*, 2020). Indeed, past studies point towards the increasing professionalization of the Airbnb platform (Farmaki *et al.*, 2020), with scholars questioning whether it may even be included under the sharing economy umbrella (e.g. Crommelin *et al.*, 2018). Specifically, Martin (2016) has heralded the expansion of the platform as a form of neoliberal capitalism in that the sharing practice is essentially driven by economic self-interest rather than sharing motivations. Regarding this view, Airbnb commercializes aspects of urban life previously beyond the reach of the market, inflicting problems such as tourists' antisocial behaviour, changes in neighbourhood character, and loss of affordable housing (Gutiérrez *et al.*, 2017; Richards *et al.*, 2019). As such, Airbnb acts as an exogenous driver of negative impacts that potentially threaten the well-being and resilience of local communities (Volgger *et al.*, 2018).

According to Adger (2000, p. 361), community resilience refers to 'the ability of communities to withstand external shocks to their social infrastructure'. It follows logically that community resilience is likely to involve the ability of local residents to absorb negative impacts and adaptively respond to threats (Norris *et al.*, 2008). For example, the negative externalities caused by Airbnb have led to an increasing number of protests from residents in several cities (Hassanli *et al.*, 2019). Accordingly, of relevance here is community resilience to Airbnb-induced transformations. Particularly, emphasis is placed on the extent and the way in which local residents exercise agency at developing coping and control mechanisms. Arguably, resilience and resident disturbance may range considerably, resulting in variation in local residents' adaptive responses (Cheer and Lew, 2018).

Even so, the perceptions and responses of local residents remain largely overlooked in tourism scholarship. This omission is surprising as the proliferation of Airbnb in residential areas, and the subsequent opposition it may yield from residents, is likely to result in unwanted resident–tourist interactions that may negatively affect both local communities and the tourist experience (Gurran and Phibbs, 2017). Therefore, the perceptions of residents living in areas where there is a high concentration of Airbnb rentals need to be investigated (Ioannides *et al.*, 2019) if the phenomenon is to be regulated in a sustainable way to safeguard community well-being. To this end, this chapter examines the perceptions of the residents of the Koukaki area (Athens, Greece) which, like many other residential areas located close to historic city centres, is experiencing unprecedented growth in Airbnb rentals. According to Coldwell (2016), the area saw an 800% increase in Airbnb listings over the last few years as it transformed from a quiet residential area to a neighbourhood popular among Airbnb guests. In so doing, we employ social exchange theory (SET) to illuminate understanding of how residents' perceptions are formed in relation to their exchange relationship with Airbnb guests.

The rest of the chapter is organized as follows. First, we present the literature on the impacts of P2P accommodation on neighbourhoods before explaining the SET. Then, we explain and justify the methodology guiding the study. Subsequently, we present and discuss study findings. The implications arising from this study in relation to community resilience are drawn as conclusions.

The Invasion of Airbnb in Neighbourhoods

The Airbnb platform has grown exponentially, particularly in urban areas where it has been recognized as an instigator of tourism expansion (Ioannides *et al.*, 2019), exerting profound effects on local communities. As tourists began to stay in residential zones, neighbourhoods with a large supply of Airbnb rentals have particularly experienced the impact of Airbnb's growth (Stors and Baltes, 2018). As Gyódi (2019) argues, the popularity of Airbnb is responsible for creating new tourist spaces that are reflective of the overtourism phenomenon. Within this context, an array of problems has begun to emerge in the newly touristified neighbourhoods, including increased noise, traffic and pollution, a rise in property prices and rent, a loss of neighbourhood identity and an increased sense of insecurity (Gurran and Phibbs, 2017; Richards *et al.*, 2019). As a result, the residents of neighbourhoods impacted by the growth of Airbnb have expressed resentment against tourists staying in Airbnb properties (Colomb and Novy, 2017). This has been evidenced through an increasing number of anti-Airbnb protests and demonstrations organized in several cities across the world (*The Economist*, 2018). Indeed, there are several reports indicating conflict between residents and Airbnb guests arising due to the latter's unruly behaviour (Guttentag, 2015).

Although such conflictual behaviours are common in neighbourhoods, the surge of Airbnb rentals in residential areas seems to have aggravated existing neighbour problems. Specifically, the expansion of the phenomenon has added another layer of complexity as residential neighbourhoods are rapidly being transformed into tourist spaces. While studies on Airbnb have previously touched upon several key issues (e.g. Amore *et al.*, 2020; Caldicott *et al.*, 2020), the specific impacts in neighbourhood settings as perceived by residents have been largely overlooked by tourism and hospitality scholars. This study thus contributes to the extant literature by offering insights into the perceptions of residents of the impacts of Airbnb on their neighbourhoods. The study adopts SET in its examination, which has been widely utilized within tourism studies to examine the resident–tourist relationship. It does so by explaining why and under which conditions residents respond to tourism impacts the way they do (Ap, 1992). Despite the fact that SET was introduced several decades ago, it remains a powerful force in tourism research. In fact, several recent studies corroborate its utility as a general analytic approach to a wide array of tourism exchanges (see Hadineyad *et al.*, 2019), including resident–guest interactions in P2P accommodation (e.g. Tussyadiah, 2016; Priporas *et al.*, 2017). As such, SET is proposed as a suitable overarching theory to aid us in better understanding the sharing economy in hospitality and tourism contexts (Altinay and Taheri, 2019).

Social Exchange Theory (SET)

SET is a popular paradigm within tourism studies that aims to examine the social behaviour of people and/or groups (Ap, 1992) by focusing on their perceptions of the costs and benefits of relationships. According to Sharpley (2014), SET has been widely adopted within the tourism research domain as it enables the examination of residents' perceptions of the socio-economic and environmental impacts of their exchanges with tourists. Generally speaking, residents partaking in an exchange relationship with tourists will evaluate the relationship more favourably if the benefits received from the interaction outweigh the costs (Ap, 1992). For example, if the residents of a destination perceive that tourism offers economic, social and psychological benefits in the form of revenue, job opportunities and sociopsychological well-being, then they will most likely support tourism development and show friendly behaviour towards tourists. On the contrary, if the residents of a destination feel that tourism yields more negative impacts than positive ones, they will most likely perceive the encounter with tourists unfavourably, exhibiting either indifference towards tourists or hostility.

Overall, the exchange relationship between tourists and residents is considered sequentially as a process consisting of three stages: (i) the pre-exchange stage; (ii) the actual exchange; and (iii) the post-exchange stage. The first stage is important as it may emerge voluntarily due to the benefits accrued by the local community or it may be brought about as an imposition by a third party (e.g. the government). The second stage of the exchange process consists of three interconnected components (antecedents, the exchange relation and the form of the exchange relation) that interact to influence perceptions. For instance, if the antecedents (also known as opportunities) are perceived by one of the parties as rewarding, fair and equitable then the exchange usually takes place (Moyle *et al.*, 2011) and vice versa. None the less, the relationship may be imposed by one actor on another (Ap, 1992). For example, tourism may be imposed on certain community members who are unable to avoid tourists. The exchange relationship may thus emerge as a function of power between two parties which may be balanced or unbalanced. The notion of power as a means of understanding the resident–tourist exchange situation is addressed by Kayat (2002, p. 175), who notes that 'one's power is derived from control over the resources that the partner needs and from control over the returns obtained from the exchange'. Balanced exchanges result from transactions between those with equal control over resources. Exchange transactions between those who have high and low levels of power are described as unbalanced. The last stage of the process comprises the exchange transaction evaluation and the consequences of the exchange, which occurs once the exchange is complete. At this stage, each actor evaluates the actual exchange and associated effects in order to decide if the exchange behaviour will cease, in the case of negative impacts, or continue in the case of positive impacts. However, even in the case of a negative evaluation, the relationship may not be terminated as the exchange may be imposed (Ap, 1992) or perceived as necessary (Lindberg *et al.*, 2001).

Within tourism research, SET has been utilized primarily in quantitative analyses (Dann *et al.*, 2019) focusing on variables that influence residents' perceptions of their exchange relationship with tourists. As such, in-depth investigations of the factors shaping these perceptions are lacking (Sharpley, 2014). This study follows, therefore, a qualitative approach in examining residents' perceptions of the impacts of Airbnb in their neighbourhoods through the lens of SET. The following section explains the methodology guiding the data collection and analysis.

Methodology

We adopted an exploratory research approach, with the aim to examine in-depth the conditions under which residents in the area of Koukaki (Athens) develop positive or negative perceptions with regard to the impacts of Airbnb on their neighbourhoods. Specifically, we performed semi-structured interviews with the questions being framed according to the stages of the exchange relationship, as suggested by SET. Therefore, among the questions asked were the following: (i) When did you become aware that properties in the area were transformed into Airbnb rentals?; (ii) What were your expectations from Airbnb guests?; (iii) How would you describe your interaction with guests?; (iv) What benefits/costs did Airbnb bring into the neighbourhood?; and (v) How did you deal with these?

We used purposive and snowball sampling to target participants. In the first instance, purposive sampling was used to select participants based on the following criteria: (i) participants had to live permanently in the area of Koukaki; and (ii) they had to have experience with Airbnb guests within the setting of their neighbourhood. Purposive sampling allows researchers to use their a priori theoretical understanding of the topic (Robinson, 2014) and select participants deemed knowledgeable about the study (Schutt, 2018). Overall, 14 participants were identified by a member of the research team who lives in the area of Koukaki (marked in italics in Table 2.1). After the initial participants were selected, we adopted snowball sampling by asking participants to refer other suitable subjects to the researcher (Noy, 2008). In the end, a total of 51 residents were included in the sample. Care was undertaken to ensure that adequate diversity characterized the sample (Kim and Ritchie, 2014) in terms of age and gender. Table 2.1 presents the characteristics of study participants.

The interviews took place between November 2017 and March 2018 and were conducted face to face and on a one-to-one basis with a member of the research team. The interviews lasted on average 45 minutes and, before each interview, the researcher explained to the participant the purpose of the study, ensuring him/her of anonymity and confidentiality issues (Fontana and Frey, 2005). The interviews were performed in Greek and were translated by a professional translator. After checking the transcripts for accuracy, data were analysed using thematic analysis (Braun and Clarke, 2006). In particular, transcripts were read several times by the researchers to familiarize themselves with

Table 2.1. Profile of the interviewees.[a]

Interviewee number	Gender	Age (years)	Occupation
1	*Female*	*42*	*Civil servant*
2	*Male*	*71*	*Retired*
3	*Male*	*62*	*Bookseller*
4	*Male*	*44*	*Accountant*
5	*Female*	*68*	*Retired*
6	*Female*	*39*	*Physiotherapist*
7	*Female*	*32*	*School teacher*
8	*Female*	*29*	*Waitress*
9	*Male*	*49*	*Newsagent*
10	*Female*	*30*	*Make-up artist*
11	*Male*	*54*	*Dentist*
12	*Male*	*36*	*Telephone salesperson*
13	*Male*	*49*	*Taxi driver*
14	*Male*	*50*	*Bank cashier*
15	Female	66	Retired
16	Female	35	English teacher
17	Male	42	Electronics technician
18	Female	70	Retired
19	Male	54	Medical representative
20	Female	52	Optician
21	Male	31	Web developer
22	Female	39	Clothing-shop assistant
23	Male	39	Insurance sales agent
24	Female	38	Housewife
25	Female	44	Unemployed
26	Female	56	Electrical engineer
27	Male	68	Retired
28	Male	40	Unemployed
29	Female	67	Retired
30	Male	39	Car salesman
31	Female	51	Telephone operator
32	Male	45	Grocer
33	Male	55	Mechanic
34	Female	52	Hair-salon owner
35	Female	70	Retired
36	Male	55	Civil servant
37	Female	35	Unemployed
38	Male	50	Mini-market owner
39	Female	45	Housewife
40	Male	35	Furniture salesman
41	Female	40	Seamstress
42	Male	32	Data entry clerk
43	Male	51	Pharmacist
44	Female	48	Clothing-shop owner
45	Male	58	Locksmith
46	Female	33	Confectioner
47	Female	74	Retired
48	Female	40	Housewife
49	Female	45	Nurse
50	Female	50	Pharmacist
51	Male	37	Musician

[a]Interviewees in italics are the initial 14 participants identified by a member of the research team who lives in the area of Koukaki.

the data and identify patterns of themes. Blocks of text were copied verbatim, reorganized and cross-referenced to allow the identification of thematic categories. Subcategories also emerged which were combined with pre-identified themes to allow for greater elaboration on key issues (Hennink *et al.*, 2010).

Findings

Pre-exchange stage

In the first instance, we asked participants to elaborate on how they became aware of Airbnb rentals in their neighbourhood. The majority of participants stated that they realized properties were converted into Airbnb rentals when they heard noise from renovations in neighbouring apartment blocks. These constructions appear to have disrupted neighbourhood life as a result of excessive debris and dust as well as the constant noise of drilling. In this regard, participants argued that 'no one asked or informed residents of these renovations' (male, 44, accountant) leaving them feeling that the relationship with Airbnb guests was imposed on them against their wishes. In addition, participants commented on finding out about the transformation of neighbouring properties into Airbnb rentals through their neighbours themselves, who were evicted from the properties by landlords wishing to invest in short-term rentals. 'Our neighbours have been displaced ... after being evicted they tried to find another apartment in Koukaki but the rent increased so much because of Airbnb, they couldn't find anything' cried a resident (female, 38, housewife). In relation to this point, one participant commented on how a construction company has bought several apartments in his building with the aim of converting them into short-term rentals on P2P-accommodation platforms. In his words, 'my home is becoming a hotel' (male, 71, retired). In some cases, residents found out about neighbouring properties being turned into Airbnbs when they unexpectedly encountered foreigners in their apartment blocks. Within this context, a common concern expressed by participants was how their neighbourhoods were losing their identity. As a resident said 'a neighbourhood is its people, when they are gone so is the neighbourhood' (female, 32, teacher).

The actual exchange

Subsequently, we asked participants to share their opinions regarding the antecedents of the exchange relationship with Airbnb guests and the experiences with them. The majority of participants expressed negative perceptions, with unfavourable expectations from the presence of Airbnb guests in the area emerging as a determining factor. 'I prefer to live next to people I know, not strangers from another country that I cannot communicate with' said a participant (female, 74, retired). On a similar note, other participants identified the lack of economic or social benefits as a reason for the unfavourable perceptions towards Airbnb guests. One participant mentioned: 'the professionals get money from tourists, but locals don't get anything so why have them on my doorstep?' (male, 51, pharmacist). Only some participants disagreed, illustrating more positive perceptions towards the influx of Airbnb guests in their neighbourhood, highlighting that in the case of problems these will be transitional as 'guests come and go' (male, 49, taxi driver). These participants, in particular, argued that visitors may bring life to the area as well as economic gains as 'people have the opportunity to rent their spare properties to short-term travellers to make some extra money' (female, 39, physiotherapist) or 'tourists will buy things from the local shops' (female, 48, shop owner). It thus appeared that participants with a more positive stance towards Airbnb guests expect positive outcomes from the exchange relationship. Interestingly, positive views over the issue of Airbnb came from respondents who were relatives of Airbnb hosts and those informants owning shops in the area. This corroborates well with the robust finding of SET that those, directly or indirectly, benefitting from visitor spending have more positive perceptions of tourism development (Kayat, 2002).

Despite the dominance of negative expectations, the exchange relationship between residents and Airbnb guests actually formed, albeit it was involuntary and rather unbalanced in character. In relation to this point, participants highlighted their lack of control over the situation, criticizing property owners as 'acting irresponsibly towards residents' (female, 45, nurse) while identifying the lack of a legislative framework as conducive to the growth of Airbnb rentals.

In this context, participants began to recognize a variety of problems emerging from their encounters with Airbnb guests, including excessive noise and litter, damage to properties and communal areas of buildings (i.e. elevators) as well as physical confrontations with guests as a result of their antisocial behaviour. As a participant put it 'tourists drink, party and make too much noise without a consideration for the people living in the neighbourhood, without thinking that there are young children living here' (female, 68, retired). Other participants agreed, arguing that the social fabric of the neighbourhood has been transforming since the growth in Airbnb rentals, leading to 'a change in neighbourhood character' (male, 54, medical representative).

In this context, participants pointed towards the role of property owners and Airbnb hosts, many of whom appear to show indifference to the complaints of residents and fail to undertake any actions to safeguard the well-being of the local community. In the words of a participant 'many hosts are not even permanent residents, so they don't care about the neighbourhood' (female, 56, electrical engineer). In addition, some participants reported calling the police to control unruly tourists, with officers being unable/unwilling to report complaints, instead pressurizing them to privately find a compromise with guests and/or hosts. As a participant commented, 'the police said they cannot do anything because the guests are not the owners of the properties' (male, 37, musician).

Post-exchange stage

Evidently, the majority of participants shared predominantly negative perceptions over Airbnb's presence in their neighbourhood, with their evaluation of associated outcomes of the exchange relationship being equally negative. The evaluation of participants was mostly shaped by personal impacts and/or wider problems pertinent to the neighbourhood level. For instance, several participants acknowledged that they feel anxiety and irritation from the situation which is 'threatening the mental and physical well-being' of locals (male, 39, car salesman). In particular, concerns over the neighbourhood's transformation seemed to emerge from older residents who expressed fear and higher levels of stress as a result of the influx of tourists in the area, with whom many residents cannot communicate due to a lack of language knowledge. Furthermore, participants acknowledged a range of neighbourhood problems that emerge from the growth of Airbnb, such as higher rents and council taxes among others.

Consequently, many participants expressed a desire to move to another area. Yet, financial and personal factors were inhibiting them from doing so. As a participant stated 'my children go to school here and my work is close by, I cannot just up and leave' (female, 33, confectioner). Therefore, a form of resigned acceptance was evident among a large majority of the local community as they felt powerless in controlling the situation. In the words of a participant 'the situation is not in our hands' (female, 52, optician). Despite such feelings, several participants described acts of resistance that they undertake, either actively or passively, in order to assume control and mitigate the problems. For example, several participants admitted to opposing Airbnb by refusing to use it as guests. Others attempted to seek legal advice to find out their options while a few residents appear to have become self-organized, seeking signatures from neighbours to petition formally against Airbnb. These participants expressed determination in 'driving Airbnb out' (female, 45, nurse) of the neighbourhoods. Indeed, such sentiments seem to increase among Athenian residents as recently observed in a large anti-Airbnb protest that took place in the city in late autumn 2019 (Fig. 2.1).

By contrast, there were few participants who acknowledged that some benefits may emerge from Airbnb rentals, which they mostly translate into 'financial gains for local shop owners' (male, 50, mini-market owner). While most of these participants showed a passive supportive stance towards Airbnb, some contemplated becoming Airbnb hosts themselves arguing that 'this phenomenon has proven to be profitable for hosts ... why not see this as an opportunity and exploit the fruits it may give' (male, 31, web developer).

Discussion and Conclusion

The growth of P2P accommodation has been exponential, with many tourism destinations witnessing changes in urban life and tourist

Fig. 2.1. Graffiti drawn in Exarcheia (Athens) following an anti-Airbnb protest. Photo courtesy of Euronews Greece.

guests proving disruptive for local residents as they spread beyond conventional tourism spaces into residential ones (Ioannides *et al.*, 2019). In this context, the present study used SET to delve into residents' perceptions of the impacts of P2P accommodation in their neighbourhood, as a result of their interactions with Airbnb guests. Study findings show that residents identified a range of economic, social and environmental P2P-accommodation impacts, which they perceived as both highly intrusive into their daily life and a cause of neighbourhood problems. Indeed, the majority of research participants in this study emphasized that imposed costs outweighed any positive outcomes brought about by Airbnb and their interactions with guests. As such, the rise of P2P accommodation contributed to residents' sense of touristification, transforming their neighbourhood and residential environment into a tourism space with a constant rotation of visitors and becoming disruptive to its social fabric.

While the presence of tourists in residential areas is not uncommon, P2P-accommodation rentals increase the concentration of tourists and length of stay, thus bringing an acceleration of gentrification and associated negative effects – such as crowding and disruptive behaviour (Gurran, 2018). In residential areas, however, tourism impacts are particularly concerning as they occur at the micro level of neighbourhoods, which represents the basis of social structures and social life (Chesire and Fitzgerald, 2015). The displacement of residents through evictions, high rental rates, and the weakening of a local sense of community and social ties as guests replace long-term neighbours are clearly suggestive of the disruptive effects of P2P accommodation on residents' social well-being. Unsurprisingly, residents in this study generally developed negative perceptions towards this form of development, which were further aggravated by the irresponsible behaviour of guests and hosts, who were seen as having no long-term vested interest in the community. Ultimately, as indicated by study participants, the impacts experienced may incur a range of health-related and psychological problems which, according to Janusz *et al.* (2017), put community resilience at stake.

As MacCrea *et al.* (2016) argue, relationships between resilience and community well-being become salient when the latter is undermined by a challenge facing a community. They go on to explain that 'community resilience is focused on improving community well-being ... which emphasizes the active nature of community resilience' (MacCrea *et al.*, 2016, p. 196). In turn, this active nature of community resilience suggests that it entails community actions to adapt to the changing conditions (Norris *et al.*, 2008). From a community resilience perspective, then, and in response to their negative perceptions, residents adopted varying types of behaviour to adapt and respond to the adverse effects of P2P-accommodation growth in their neighbourhood. These included anti-Airbnb activism and confrontational behaviour towards guests. In the absence of stakeholders (such as the local police) taking responsibility, as in the case of the largely unregulated P2P-accommodation sector (Farmaki *et al.*, 2019), residents are likely to take mitigating action to address emerging problems. Perceptions of powerlessness to control the situation, further fuel their feelings of disappointment and frustration. Other respondents demonstrated a kind of 'passive receptivity' of the new context in which they live, still expressing their unwillingness to engage in exchange relationships with Airbnb guests. Considering that value co-creation in the context of P2P accommodation is largely derived from the quality of guest–resident interactions (Johnson and Neuhofer, 2017), potential conflicts between the two parties, or residents avoiding such interactions altogether, may prove detrimental to tourist experience and satisfaction. In fact, as argued by Lin *et al.* (2017), residents' dissatisfaction with tourism development has a negative effect on their intention to co-create value with tourists. On this view, residents are particularly important in the transformation of neighbourhood settings into tourist spaces.

To capture the dynamic environment in which residents' perceptions of P2P accommodation are formed and shaped, we offer a conceptual framework (Fig. 2.2) which illustrates the formation of residents' perceptions of the impacts of P2P accommodation on neighbourhoods, highlighting the influences of key actors on residents' perceptions and behaviours. Informed by both study findings and extant literature, Fig. 2.2 depicts the growth of P2P accommodation being driven by the promise of economic and social benefits it offers to both hosts and guests (Belarmino *et al.*, 2019). Nevertheless, the growth of P2P accommodation is simultaneously fuelled by the unregulated structure characterizing the platform environment, where everyone can participate. As shown in this study, an increasing number of property owners are interested in participating in P2P accommodation and are, thus, evicting tenants in order to renovate their properties and transform them into Airbnb rentals. Using SET terminology, it could therefore be argued that residents find themselves in an involuntary exchange with Airbnb guests, which is largely unbalanced in character because residents have either low levels of power or no power at all to influence the form of the exchange. In this connection, the role of property owners and/or P2P-accommodation hosts in shaping the exchange between residents and guests is also important as is the mediating effect of the platform (Airbnb) on property owners' behaviour. Within this context of complexity, a series of positive and negative impacts have been identified in this study as originating from the growth of Airbnb in residential areas. Correspondingly, residents' perceptions may range from favourable to unfavourable. In turn,

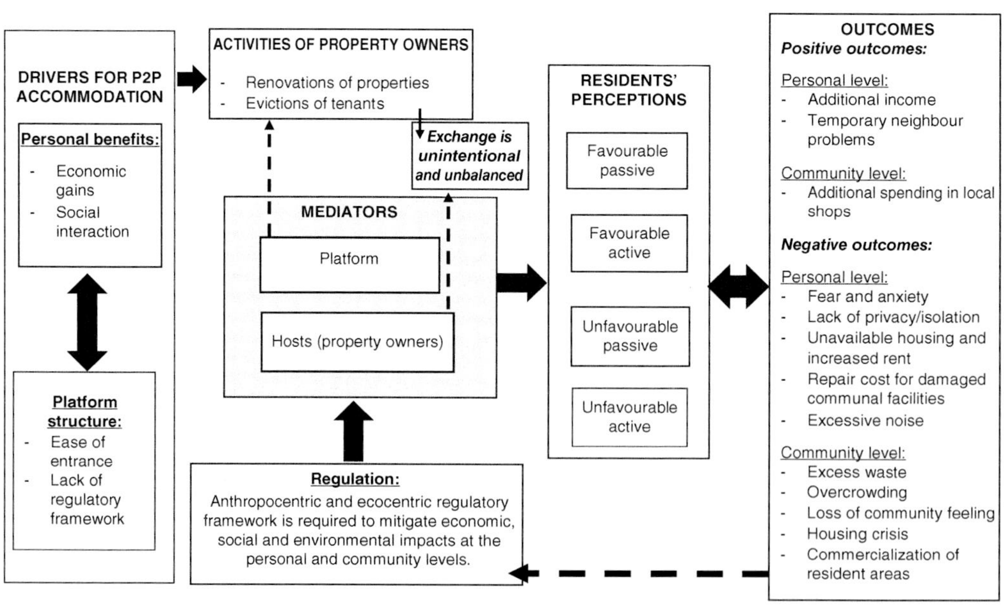

Fig. 2.2. Conceptual framework of residents' perceptions of peer-to-peer (P2P) accommodation impacts.

these perceptions lead to reactions on the part of residents who may adopt a variety of passive or active forms of behaviour to adjust themselves to the new conditions.

Conclusively, P2P accommodation plays a crucial role in transforming localities economically and socially. From a policy perspective this suggests a need for a more anthropocentric regulatory framework which considers the impacts arising from P2P-accommodation growth on neighbourhoods, at both the individual and the community levels, to stand in line with sustainability concerns. As Janusz *et al.* (2017, p. 128) note, identifying residents' main sources of concern is critical to 'design policies to match the needs of residents, build a tourism-resilient community and, as a consequence, minimize the risk of potential conflicts'. Figure 2.2 provides an early, yet informative, framework for mapping this terrain. Notwithstanding, there needs to be a consideration of the associated impacts on different stakeholder groups within the local community if appropriate mechanisms are to be adopted in order to promote resident well-being and resilience. Related to this argument is also the role of the platform and hosts/property owners in managing the exchange between residents and P2P-accommodation guests. In view of this role, embedding the needs of, and impacts on, residents within existing platform policies and regulations that shape host and guest conduct might be a worthwhile policy direction. Given the heterogeneity of residents (Jordan and Moore, 2018) and their interactions with P2P-accommodation guests, the engagement of residents in policy making remains, of course, challenging. However, by revealing some of the complexity underlying the relationships and influences in P2P-accommodation, it is hoped that this study provides a useful starting point and insights towards the inclusion of residents in policy making. Also it is hoped that the issues raised here add new knowledge to this area of research and that this promotes further studies within this field.

References

Adger, W.N. (2000) Social and ecological resilience: are they related? *Progress in Human Geography* 24(3), 347–364.

Airbnb (2020) About us. *Airbnb News*. Available at: https://news.airbnb.com/about-us/ (accessed 6 August 2020).

Altinay, L. and Taheri, B. (2019) Emerging themes and theories in the sharing economy: a critical note for hospitality and tourism. *International Journal of Contemporary Hospitality Management* 31(1), 180–193.

Amore, A., de Bernardi, C. and Arvanitis, P. (2020) The impacts of Airbnb in Athens, Lisbon and Milan: a rent gap theory perspective. *Current Issues in Tourism*, 1–14.

Ap, J. (1992) Residents perceptions on tourism impacts. *Annals of Tourism Research* 19(4), 665–690.

Belarmino, A., Whalen, E., Koh, Y. and Bowen, J.T. (2019) Comparing guests' key attributes of peer-to-peer accommodations and hotels: mixed-methods approach. *Current Issues in Tourism* 22(1), 1–7.

Braun, V. and Clarke, V. (2006) Using thematic analysis in psychology. *Qualitative Research in Psychology* 3(2), 77–101.

Caldicott, R.W., von der Heidt, T., Scherrer, P., Muschter, S. and Canosa, A. (2020) Airbnb – exploring triple bottom line impacts on community. *International Journal of Culture, Tourism and Hospitality Research* 14(2), 205–223.

Cheer, J.M. and Lew, A.A. (2018) Understanding tourism resilience: adapting to social, political, and economic change. In: Cheer, J.M. and Lew, A.A. (eds) *Tourism, Resilience and Sustainability: Adapting to Social, Political and Economic Change*. Routledge, London, pp. 3–17.

Cheng, M., Chen, G., Wiedmann, T., Hadjikakou, M., Xu, L. and Wang, Y. (2020) The sharing economy and sustainability – assessing Airbnb's direct, indirect and induced carbon footprint in Sydney. *Journal of Sustainable Tourism* 28(8), 1083–1099.

Cheshire, L. and Fitzgerald, R. (2015) From private nuisance to criminal behavior: neighbour problems and neighbourhood context in an Australian city. *Housing Studies* 30(1), 100–122.

Coldwell, W. (2016) Airbnb: from homesharing cool to commercial giant. *The Guardian*, 18 March. Available at: https://www.theguardian.com/travel/2016/mar/18/airbnb-from-homesharing-cool-to-commercial-giant (accessed 6 August 2020).

Colomb, C. and Novy, J. (2017) *Protest and Resistance in the Tourist City*. Routledge, London.

Crommelin, L., Troy, L., Martin, C. and Pettit, C. (2018) Is Airbnb a sharing economy superstar? Evidence from five global cities. *Urban Policy and Research* 36(4), 429–444.

Dann, D., Teubner, T. and Weinhardt, C. (2019) Poster child and guinea pig – insights from a structured literature review on Airbnb. *International Journal of Contemporary Hospitality Management* 31(1), 427–473.

Dogru, T., Mody, M., Suess, C., McGinley, S. and Line, N. (2020) The Airbnb paradox: positive employment effects in the hospitality industry. *Tourism Management* 77: 104001.

Farmaki, A., Stergiou, D. and Kaniadakis, A. (2019) Self-perceptions of Airbnb hosts' responsibility: a moral identity perspective. *Journal of Sustainable Tourism*, 1–21.

Farmaki, A., Christou, P. and Saveriades, A. (2020) A Lefebvrian analysis of Airbnb space. *Annals of Tourism Research* 80: 102806.

Fontana, A. and Frey, J.H. (2005) The interview: from neutral stance to political involvement. In: Denzin, N.K. and Lincoln, Y.S. (eds) *The Sage Handbook of Qualitative Research*. Sage Publications, London, pp. 695–727.

Gurran, N. (2018) Global home-sharing, local communities and the Airbnb debate: a planning research agenda. *Planning Theory & Practice* 19(2), 298–304.

Gurran, N. and Phibbs, P. (2017) When tourists move in: how should urban planners respond to Airbnb? *Journal of the American Planning Association* 83(1), 80–92.

Gutiérrez, J., García-Palomares, J., Romanillos, G. and Salas-Olmedo, M. (2017) The eruption of Airbnb in tourist cities: comparing spatial patterns of hotels and peer-to-peer accommodation in Barcelona. *Tourism Management* 62, 278–291.

Guttentag, D. (2015) Airbnb: disruptive innovation and the rise of an informal tourism accommodation sector. *Current Issues in Tourism* 18(12), 1192–1217.

Guttentag, D. and Smith, S. (2017) Assessing Airbnb as a disruptive innovation relative to hotels: substitution and comparative performance expectations. *International Journal of Hospitality Management* 64, 1–10.

Gyódi, K. (2019) Airbnb in European cities: business as usual or true sharing economy? *Journal of Cleaner Production* 221, 536–551.

Hadineyad, A., Moyle, B.D., Scott, N., Kralj, A. and Nunkoo, R. (2019) Residents' attitudes to tourism: a review. *Tourism Review* 74(2), 150–165.

Hajibaba, H. and Dolnicar, S. (2017) Substitutable by peer-to-peer accommodation networks? (Research Note). *Annals of Tourism Research* 66, 185–188.

Hassanli, N., Small, J. and Darcy, S. (2019) The representation of Airbnb in newspapers: a critical discourse analysis. *Current Issues in Tourism*, 1–13.

Hennink, M., Hutter, I. and Bailey, A. (2010) *Qualitative Research Methods*. Sage Publications, London.

Ioannides, D., Röslmaier, M. and van der Zee, E. (2019) Airbnb as an instigator of 'tourism bubble' expansion in Utrecht's Lombok neighbourhood. *Tourism Geographies* 21(5), 822–840.

Janusz, K., Six, S. and Vanneste, D. (2017) Building tourism-resilient communities by incorporating residents' perceptions? A photo-elicitation study of tourism development in Bruges. *Journal of Tourism Futures* 3(2), 127–143.

Johnson, A.G. and Neuhofer, B. (2017) Airbnb – an exploration of value co-creation experiences in Jamaica. *International Journal of Contemporary Hospitality Management* 29(9), 2361–2376.

Jordan, E.J. and Moore, J. (2018) An in-depth exploration of residents' perceived impacts of transient vacation rentals. *Journal of Travel & Tourism Marketing* 35(1), 90–101.

Kayat, K. (2002) Power, social exchanges and tourism in Langkawi: rethinking resident perceptions. *International Journal of Tourism Research* 4, 171–191.

Kim, J.-H. and Ritchie, J.R.B. (2014) Cross-cultural validation of a memorable tourism experience scale (MTES). *Journal of Travel Research* 53(3), 323–335.

Lin, Z., Chen, Y. and Filieri, R. (2017) Resident-tourist value co-creation: the role of residents' perceived tourism impacts and life satisfaction. *Tourism Management* 61, 436–442.

Lindberg, K., Andersson, T. and Dellaert, B. (2001) Tourism development. *Annals of Tourism Research* 28(4), 1010–1030.

MacCrea, R., Walton, A. and Leonard, R. (2016) Developing a model of community wellbeing and resilience in response to change. *Social Indicators Research* 129, 195–214.

Martin, C. (2016) The sharing economy: a pathway to sustainability or a nightmarish form of neoliberal capitalism? *Ecological Economics* 121, 149–159.

Moyle, W., Venturto, L., Griffiths, S., Grimbeek, P., McAllister, M. *et al.* (2011) Factors influencing quality of life for people with dementia: a qualitative perspective. *Aging and Mental Health* 15(8), 970–977.

Norris, F.H., Stevens, S.P., Pfefferbaum, B., Wyche, K.F. and Pfefferbaum, R.L. (2008) Community resilience as a metaphor, theory, set of capacities, and strategy for disaster readiness. *American Journal of Community Psychology* 41(1–2), 127–150.

Noy, C. (2008) Sampling knowledge: the hermeneutics of snowball sampling in qualitative research. *International Journal of Social Research Methodology* 11(4), 327–344.

Paulauskaite, D., Powell, R., Coca-Stefaniak, J. and Morrison, A. (2017) Living like a local: authentic tourism experiences and the sharing economy. *International Journal of Tourism Research* 19(6), 619–628.

Priporas, C.V., Stylos, N., Rahimi, R. and Vedanthachari, L.N. (2017) Unraveling the diverse nature of service quality in a sharing economy: a social exchange theory perspective of Airbnb accommodation. *International Journal of Contemporary Hospitality Management* 29(9), 2279–2301.

Richards, S., Brown, L. and Dilettuso, A. (2019) The Airbnb phenomenon: the resident's perspective. *International Journal of Tourism Cities* 6(1), 8–26.

Robinson, R.S. (2014) Purposive sampling. In: Michalos, A.C. (ed.) *Encyclopedia of Quality of Life and Well-Being Research*. Springer, New York, pp. 5243–5245.

Schutt, R. (2018) *Investigating the Social World: the Process and Practice of Research*. Sage, Boston, Massachusetts.

Sharpley, R. (2014) Host perceptions of tourism: a review of the research. *Tourism Management* 42, 37–49.

Stors, N. and Baltes, S. (2018) Constructing urban tourism space digitally: a study of Airbnb listings in two Berlin neighborhoods. In: *Proceedings of the ACM on Human-Computer Interaction* 2(CSCW), September, Article 166, pp. 1–29.

The Economist (2018) Charlemagne: the backlash against Airbnb. *The Economist*, 19 July. Available at https://www.economist.com/europe/2018/07/19/charlemagne-the-backlash-against-airbnb (accessed 6 August 2020).

Tussyadiah, I.P. (2016) Factors of satisfaction and intention to use peer-to-peer accommodation. *International Journal of Hospitality Management* 55, 70–80.

Tussyadiah, I.P. and Pesonen, J. (2016) Impacts of peer-to-peer accommodation use on travel patterns. *Journal of Travel Research* 55(8), 1022–1040.

Volgger, M., Pforr, C. and Reiser, D. (2018) Destination resilience and the sharing economy: an exploration of the case of Airbnb. In: Innerhofer, E., Fontanari, M. and Pechlaner, H. (eds) *Destination Resilience: Challenges and Opportunities for Destination Management and Governance*. Routledge, New York, pp. 142–152.

3 Airbnb Host Responsibilities and Community Resilience: the Case of Japan

Antonios Kaniadakis,[1]* Chiga Hayakawa[2] and Anna Farmaki[3]
[1]*Brunel University London, UK;* [2]*Queen Mary University of London, UK;* [3]*Cyprus University of Technology, Cyprus*

Abstract

In Japan, the shortage of accommodation facilities has become a concern due to the increasing number of tourists. In light of the 2020 Olympic and Paralympic Games (moved to 2021 due to the coronavirus disease 2019 (COVID-19) pandemic), this problem is likely to escalate and raises the need for the authorities to deal with the temporary influx of tourists. Private lodging emerges as a possible solution to this problem and also as a way to revitalize regions and repopulate vacant houses. Considering the Private Lodging Business Act recently established in Japan, this chapter explores Airbnb hosts' perceptions of their responsibilities. This leads to an increase in our understanding of the associated implications of these hosts' practice on the resilience of the local community. Drawing from semi-structured interviews with 16 Japanese hosts, we reveal insights on hosts' perceived responsibilities in the peer-to-peer accommodation context and how hosts and platforms can contribute to community resilience.

Keywords: Airbnb, host, Japan, resilience, responsibility, community

Highlights

- Private lodging emerges as a solution to the accommodation shortage problem evidenced in Japan.
- This chapter investigates Japanese Airbnb hosts' perceptions of their responsibilities in the peer-to-peer accommodation sector.
- Findings provide insights on hosts' emerging responsibilities following the establishment of relevant regulation.
- Findings illuminate understanding on the role of platforms in building community resilience.

Introduction

Hosts represent an important component of the peer-to-peer (P2P)-accommodation model, since they act as co-facilitators of P2P transactions (Moon *et al.*, 2019), contributing to guest satisfaction and intention to reuse P2P accommodation (Sthapit and Björk, 2019). Unsurprisingly, in recent years, host perspectives have attracted increasing academic attention with the majority of relevant studies investigating host perceptions, practices, relations and experiences (e.g. Cheng *et al.*, 2020; Farmaki and Kaniadakis, 2020; Farmaki and Kladou, 2020; Wang *et al.*, 2020).

*Antonios.Kaniadakis@brunel.ac.uk

 Peer-to-peer Accommodation and Community Resilience (eds Anna Farmaki, Dimitri Ioannides and Stella Kladou)
DOI: 10.1079/9781789246605.0003

This important pool of work is informative not only of the significant role that hosts play in the guest experience (Cheng and Jin, 2019) but also of the way hosts may help to mitigate the associated impacts emerging from P2P-accommodation growth on local communities (Caldicott *et al.*, 2020).

Indeed, it has been previously acknowledged that many of the problems evidenced in the sector stem from the illegal operations of several rentals and the unethical behaviours of various parties including hosts (Farmaki *et al.*, 2019; Stergiou and Farmaki, 2019). As a result, the contribution of the P2P-accommodation sector to the sustainable development of the destination may become challenged (Cheng *et al.*, 2020) and the positive outcomes derived from the P2P transactions minimized. Within this context, the responsible behaviour of P2P-accommodation participants (i.e. hosts) emerges as key to the future operation of the sector (Sthapit, 2019) and, most importantly, to the resilience capacity of the P2P-accommodation system (Leung *et al.*, 2019) and of the local communities impacted by its growth. Despite the burgeoning amount of work on P2P-accommodation hosts, little is known of their responsibilities in the sector (Farmaki *et al.*, 2019). The perceptions of hosts on their responsibilities in P2P-accommodation transactions are important in light of the formulation of regulation on the sector by countries, which often calls for the compliance of hosts to newly established legal frameworks (Nieuwland and van Melik, 2018).

Thus, the aim of this chapter is, to examine P2P-accommodation host perceptions of their responsibilities in the sector. In so doing, we focus on Airbnb hosts in Japan whose numbers, until the outbreak of coronavirus disease 2019 (COVID-19), had risen sharply. This was partly because of the popularity of P2P accommodation, which emerged as a potential solution to the country's shortage of formal accommodation facilities. In 2018, Japan's Private Lodging Business Act came into effect (Airbnb, 2018) in an attempt to provide a legal framework of operation. None the less, the requirement to register their listings in exchange for a fee (*The Mainichi*, 2018) forced many hosts to exit P2P-accommodation platforms. Those individuals who decided to continue hosting had to adapt their practices accordingly. For example, the new law called for a cap time of 180 days a year while many hosts also faced several local regulations such as ones imposing restrictions on the number of months or days they are allowed to operate (Sugiura, 2019).

Although the aforementioned regulations led to around 80% of listings being removed from the Airbnb platform, industry analysts expressed confidence that there is room for P2P accommodation in Japan, especially in rural areas where there are many unoccupied houses due to depopulation (*The Mainichi*, 2018). Specifically, P2P accommodation may provide the opportunity for property owners in rural areas to offer an authentic, local experience to tourists while also benefitting economically from using their unoccupied houses. This, in turn, strengthens the local community's resilience capacity in rural regions. Another reason for the growing potential of the P2P-accommodation sector in Japan is the country's status as the host of the 2020 Olympic and Paralympic Games (moved to 2021 due to the COVID-19 pandemic) which led to Airbnb lobbying the government to ease the restrictions imposed on short-term rentals. This followed the partnership between Airbnb and the International Olympic Committee to provide accommodation in the cities that are due to host the Games (Fujita, 2019). In particular, the Japanese local community may capitalize from the Olympic Games and strengthen its resilience through the offering of accommodation in locals' homes. P2P accommodation presents a more environmentally sustainable and socially inclusive accommodation option, which contributes to community resilience. In fact, urban resilience is one of the key criteria of the International Olympic Committee for selecting host cities. Correspondingly, this study fleshes out significant theoretical and practical implications that may illuminate existing knowledge of the role of hosts in contributing to local community resilience.

Theory

Community resilience

Community resilience is an integral aspect of sustainable community development (Bec *et al.*, 2018). Specifically, community resilience has been identified as an indicator of social sustainability that enables local communities to withstand change in society and nature (Magis, 2010). According to Magis (2010, p. 401) 'resilience pertains to the ability of a system to sustain itself through change via adaptation and occasional transformation'. Thus, community resilience emerges through the ability of the local community to develop in a continuously fluctuating environment. In this regard, community resilience has begun to dominate tourism planning and development discourse beyond narratives revolving around short-term disasters and crises, with studies emphasizing the need for host communities to develop and adjust to socio-ecological changes (Lew, 2014; Bec *et al.*, 2016; Cheer *et al.*, 2019) brought about by the constant growth of global tourism.

Within this context, scholars advocate an integrated approach to community resilience that includes all the interested parties involved in tourism planning and development. Indeed, given the complex structure of tourism and multifaceted interests of the various stakeholders involved in decision-making processes (Farmaki, 2015), a cohesive approach in managing the impacts of tourism on host communities is necessary while promoting shared responsibility (Hall and Lew, 2009). In relation to this point, it is important to define who constitutes a host community. Overall, tourism research recognizes five key stakeholder categories that are important for integrated tourism planning: (i) tourism/hospitality enterprises; (ii) public sector, government and municipal organizations; (iii) the host community; (iv) tourists; and (v) supportive sectors (Farmaki, 2019). In terms of the P2P-accommodation sector, Caldicott *et al.* (2020) identify traditional accommodation providers, P2P-accommodation host residents, non-P2P-accommodation host residents and the local government as the key stakeholders making up the host community. Evidently, P2P-accommodation hosts constitute an important component of the host community of a destination that shares an equal responsibility to that of other stakeholders in developing community resilience. Hosts can help the community to enjoy the socio-economic benefits offered by the P2P-accommodation sector by minimizing relevant negative impacts (Stergiou and Farmaki, 2019).

Berkes and Ross (2013) argued that community resilience can be developed through agency and self-organization. P2P-accommodation hosts have emerged as a 'community of practice' (Wenger, 1998) that shares knowledge and experience while also setting boundaries between what might be understood as responsible hosting behaviour. For instance, there are numerous online host forums and groups that facilitate knowledge sharing among hosts; this knowledge, in turn, becomes institutional practice within the P2P accommodation sector (Farmaki and Kaniadakis, 2020). Therefore, the responsible behaviour of P2P-accommodation hosts as a subcomponent of the host community is a significant indicator and contributor of the resilience capacity of the local community. None the less, little is known of hosts' responsibility in P2P accommodation settings despite indications of immoral behaviour by hosts which not only threaten community resilience but also challenge the ability of P2P accommodation to support the sharing philosophy of the P2P economy that allegedly promotes sustainability (Midgett *et al.*, 2018; Liu and Chen, 2020).

P2P-accommodation host responsibility

Generally speaking, responsibility is defined as the compliance to rules and the caring of duties as well as the unforceful application of certain values as evidenced in people's actions and motives (Selznick, 2002). Responsibility may be personal or collective. While the first involves individuals that take responsibility in an active way (Linley and Maltby, 2009), the latter is concerned with the responsibilities that groups (e.g. companies) have. Overall, collective responsibility is more difficult to assign as it involves several individuals and/or groups of people (Kaufman, 2015) that may often diffuse responsibilities to others. This may be the case when too many stakeholders are involved and responsibility is not clearly assigned (McGregor, 2017). Correspondingly, it is possible that external factors inhibit the assumption of responsibility by restricting capacity to act responsibly (Middlemiss, 2010). In addition to personal and collective responsibility, it may be possible to distinguish between individual and institutional responsibility (Green, 2002) wherein institutional agents appear to have greater resources in regulating a problem and influencing others' behaviour.

In relation to P2P-accommodation settings, institutional responsibility is largely undertaken by platforms as their policies and platform design seem to guide host and guest behaviours. For example, Airbnb's introduction of superhost and superguest badges as well as the Airbnb Plus and Airbnb Luxe initiatives (allowing listings to be rated according to a set list of criteria) highlight the influence the platform exerts on hosts (Farmaki and Kaniadakis, 2020). Likewise, the platform's mutual review system, data protection laws and adherence to social responsibility, as evidenced by their recent move to remove profile photos before the booking is finalized (Farmaki and Kladou, 2020), stand as examples of institutional responsibility (Helberger *et al.*, 2018). Moreover, government attempts to establish a regulatory framework for the P2P-accommodation sector emerge as another way of exercising institutional responsibility.

However, while in some places the P2P-accommodation sector is subject to a widely diversified regulatory framework, others lack this. This makes individual responsibility (either personal or collective as evidenced through host forums/groups) a more appropriate self-organization form towards the emergence of community resilience. Indeed, the notion of the sharing economy, to which P2P accommodation belongs, emphasizes personal responsibility of the individuals involved in P2P transactions (Sundararajan, 2016). To this end, we seek to examine the perceptions and responsibilities of Airbnb hosts in the P2P-accommodation sector. The chapter focuses on Airbnb hosts in Japan, which provides an interesting cultural context for examining host self-perceived responsibilities since the growth of the P2P-accommodation sector was disrupted in 2018 by the introduction of a new law on private lodging. This led to 80% of the platform's listings being wiped out from the platform.

Methodology

We adopted a qualitative approach since we regard this as more appropriate for the purpose of examining hosts' self-perceived responsibilities. Specifically, semi-structured interviews were performed with Airbnb hosts in Japan. We selected the sample of hosts using a purposive approach to ensure that particular types of cases were represented in the final sample (Robinson, 2014). As such, the sample was drawn from the population of Airbnb hosts; rather than targeting professional hosts we aimed to include in the sample ordinary people who provide their unused accommodation spaces to guests. Although there are various types of hosts on P2P accommodation networks such as professional hosts managing multiple listings and hosts sharing a room in their house (Farmaki and Kaniadakis, 2020), for the purposes of this study we opted to focus on individual, non-professional hosts. None the less, we ensured enough diversity was present in the sample in terms of age, gender and host background (Ritchie *et al.*, 2013). This maximized the validity of the findings. For example, interviewees were based in various locations (e.g. eight prefectures in the Tokyo metropolitan area, Tokyo suburbs and Kyoto).

The interviewees were targeted using online Airbnb host forums whereby a member of the research team who is Japanese posted an open call, asking hosts to participate in the study while ensuring their anonymity would be maintained. In total, 16 people agreed to participate in the study (female = ten, male = six) out of which five had only one listing on the platform while eight had two to three listings and three had more than three listings. The listings included the hosts' own home as well as second homes not used regularly. With the exception of two hosts who used salaried staff to run their property rentals, the rest managed (i.e. cleaning) their listings on their own, with six hosts sharing the space with their guests. The interviews were conducted between April and June 2019 on a one-to-one basis via video call through Skype, Facebook Messenger and Lin. The interviews were performed in Japanese and, subsequently, translated into English. Each interview lasted approximately 60 minutes and was digitally recorded following the interviewee's consent.

Data was analysed using thematic analysis (Braun and Clarke, 2006) to illuminate underlying themes in the discussion. Specifically, data was broken down, analysed, compared, categorized and assembled accordingly to form initial thematic categories. Following an initial round of open coding, topics emerging from the data were grouped into interrelated themes by copying, reorganizing and comparing thematic categories while refining the data under each theme to identify subcategories (Goulding, 1999). Then, subcategories were combined with the themes initially identified to validate relationships, refine and further develop thematic categories (Strauss and Corbin, 1990). As such, this allowed elaboration of the key issues (Hennink *et al.*, 2010). Overall, the three key themes emerging from the data analysis are: (i) host views of the new regulatory framework in Japan; (ii) hosts' motives to continue their hosting activity on the Airbnb platform; and (iii) platform responsibilities towards guests, hosts and community.

Findings and Discussion

Host views of the new regulatory framework

Given the adverse effect that the Private Lodging Business Act has had on P2P-accommodation growth in Japan (*The Mainichi*, 2018), we first opted to ask hosts to express their opinion on the newly established regulatory framework. While the informants acknowledged the growth potential of Airbnb in Japan, they expressed concerns over the introduction of the new law, which had disrupted the entire P2P-accommodation sector. Specifically, the informants mentioned that when the new law was introduced in 2018, Airbnb announced that it would delete listings that did not comply with the newly established regulation. Indeed, just before the new law came into effect, Airbnb deleted four-fifths of its approximately 60,000 listings in Japan (Airbnb, 2018).

Following the introduction of the Private Lodging Business Act, some hosts said that they managed to obtain permission for simple lodging (referring to inns and hotels vs private lodgings) after meeting strict equipment requirements under the Inns and Hotels Act. None the less, many of the hosts were disqualified from the private lodging category due to the 180-day restriction (Sugiura, 2019). Therefore, it is not surprising that hosts criticized the Private Lodging Business Act as being more favourable towards professional listings (hotels and inns) and less friendly towards individual property owners. In particular, concerns were expressed in relation to the new law with informants stating that the regulation was making it rather difficult for individual, non-professional hosts to operate on the Airbnb platform. Indeed, Airbnb in Japan began to promote listings of accommodation such as inns and hotels. As a result, the number of professional hosts has gradually increased while individual hosts began to exit the platform (*The Mainichi*, 2018). Such a development means that the opportunity for meaningful host–guest contact is minimized; hence, as hosts do not benefit from host–guest relations the local community's resilience capacity is minimized.

In relation to this point, hosts mentioned Airbnb's new initiative in Japan called 'Airbnb Partners', which is a partner ecosystem designed as a sharing collaboration with Japanese companies (Airbnb, 2018). The initiative aimed at including 117 companies such as real estate agents and local enterprises in this corporate ecosystem to help institutionalize professional hosting and encourage hospitality professionals to join Airbnb. Among the incentives for professional hosts to join the Airbnb platform is the extended access offered to international markets. As an interviewee (male, 36, Tokyo suburbs) explained:

> Today, two-thirds of Airbnb's listings are people who own or work for ryokans, Japanese traditional-style inns. Many inn owners, especially in rural areas, realized that they can't reach foreign guests through Japanese travel agencies, so they started to use Airbnb's platform and got bookings.
>
> (Interviewee 1)

Correspondingly, much of the criticism of informants was over the new regulatory framework and its focus on the increasing professionalization of hosting in P2P accommodation, which seems to be growing with the 'blessing' and activities of the Airbnb platform. In light of this situation, we asked informants to explain the motives behind their decision to continue hosting on the platform.

Hosts' motives to continue hosting on the Airbnb platform

Although the informants described the situation for individual, non-professional hosts in Japan as unfavourable, some hosts still choose to continue hosting. While informants acknowledged financial motives as important, other factors were recognized as equally significant in the informants' decision to host on the platform. One key characteristic of the informants, as transpired from the data analysis, is that they are internationally oriented and motivated by 'international exchanges' of culture and experiences. For instance, most of the informants have lived abroad for extended periods (e.g. for study or work), they travel frequently and speak English. As one of the interviewees (male, 64, Tokyo metropolitan area) explained: 'I have lived in foreign countries for about 20 years and many foreign people helped me. After my retirement, I decided to start this business to express my thanks to them. I enjoy international exchanges' (Interviewee 2).

Part of the 'exchanges' that the hosts talked about, is to give visitors rich experiences in Japan that may involve farming (in rural areas) and cooking. For instance, some hosts said that they even cook for their guests, taking pride in the fact that they can share aspects of their culture with foreigners. Apart from international cultural exchanges, hosts also seem to be motivated by their need to contribute through hosting to the guest experience (Cheng and Jin, 2019). This need is related to their own previous experiences as travellers and the challenges they had faced. For example, one interviewee (female, 42) commented on how she approaches hosting as a mission to support single women travelling alone: 'When I was young, I often travelled alone. But I couldn't find many inns to accept a single female traveller. Now, I run this business hoping to help single female travelers ... this is my purpose' (Interviewee 3).

In this context, informants expressed criticism of the platform's response to the Private Lodging Business Act stating that it signals a departure from the original vision of the sharing economy and home sharing as it represents a move towards more corporate models. While expressing their concerns, informants differentiated between what they do and the type of professional services that hotels and inns offer through the Airbnb platform, echoing past study findings (e.g. Farmaki and Kaniadakis, 2020). Specifically, hosts described themselves as the 'representatives' of the original Airbnb vision which was about sociality and rich international experiences. In the words of an interviewee (female, 34, Kyoto):

> There are roughly two types of host: non-professional hosts and professional hosts. The number of non-professional hosts is smaller than that of professional hosts. But it seems to me that the non-professional hosts still have the original philosophy of Airbnb which was 'Make friends in the world!' and this is how they welcome guests.
>
> (Interviewee 4)

By contrast, professional hosts – most of whom are identified by our interviewees as estate agents – were accused of listing vacant rooms for profit-making purposes. As an interviewee (female, 34, Kyoto) put it, 'for them, guests are only customers who pay money to them' (Interviewee 4). Therefore, we diverted our attention to the practices that individual, non-professional hosts undertake in relation to their perceived responsibility as members of the host community (Caldicott *et al.*, 2020).

Platform responsibilities towards guests, hosts and the community

When asked about the practices they undertake as part of hosting, informants identified a range of activities including accepting booking requests, screening guest profiles, conducting/reading reviews and generally ensuring a positive experience is offered to the guest, without any negative impacts on neighbours. None the less, informants commented on the various problems that arise in different hosting activities as a result of the platform's changing policies. For instance, hosts mentioned that although the two-way review system that is in place seems transparent and efficient, it remains one sided and leans towards benefitting guests rather than hosts. Several hosts, for example, said that often guests use the rating system to negotiate changes in the provision of hospitality services and the manner in which transactions are made. Specifically, hosts expressed a desire for the platform to adopt a fairer online rating system that supports the screening process for both hosts and guests. In the words of an informant (male, 51, Tokyo metropolitan area):

> I wish Airbnb would be responsible for maintaining fairness in reviews about the host. At this moment, all the reviews including malicious ones are open. If I get a one-star review, my guest house may be closed on the Airbnb site. When the complaints are too unreasonable, I expect Airbnb to delete the review.
>
> (Interviewee 5)

Moreover, hosts pointed out that the in case of a group of guests, only the person who made the booking is accountable while the rest are not included in the review or in the rating process. Also, informants warned that while guests do not seem to be required to post their pictures on the platform due to personal data protection regulations (Farmaki and Kladou, 2020), hosts do not enjoy the same privilege. As an interviewee (female, 39, Tokyo metropolitan area) confirmed:

> It [screening process] is obviously arranged to be disadvantageous to hosts. For example, hosts could see pictures of the guests before but now they can't. In contrast, hosts must disclose everything like pictures of their faces, pictures inside their houses, etc.
>
> (Interviewee 6)

As a result, hosts expressed feeling disadvantaged, exposed and unprotected. The informants argued that Airbnb should be more involved in the screening and background checking for guests with the same rigour that it is done to them as hosts. Specifically, an interviewee (female, 41, Kyoto) commented that:

> Airbnb must screen and audit hosts and guests honestly and sincerely. I think that the reason why guests can use a cartoon character instead of their own picture is that Airbnb doesn't check the profiles well enough.
>
> (Interviewee 7)

In relation to this point, informants identified the need for a stronger screening process in order to select the right guests as integral to the avoidance of problems with neighbours. In addition, informants said that the online rating system acts as an additional pressure on hosts to accept as many guests as possible. However, this might have implications for local neighbourhoods as tourist numbers grow, bringing with them a range of potential problems including noise and disruption (Stergiou and Farmaki, 2019). An interviewee (male, 36, Tokyo suburbs) explains:

> A new index was introduced recently as an approval rate. It is intended to get hosts to accept more guests. I explained to Airbnb Japan staff that because it's a kind of homestay, I can't approve every request, and I am responsible to my neighbours. I promised my neighbours that I would check before accepting guests.
>
> (Interviewee 1)

Hosts' contribution to community resilience

In this context, individual non-professional hosts in Japan have begun to be actively involved in 'negotiating' a fairer and more resilient governance model for private lodgings through Airbnb. This is done through an emerging 'community of practice' (Wenger, 1998), which tries to distance itself from professional hosts while promoting informal interactions and cultural exchanges. In particular, this emerging community of hosts negotiates a fairer rating system with the platform that equally supports the screening process for both hosts and guests. Also, it facilitates the sharing of know-how among hosts (Farmaki and Kaniadakis, 2020) as well as the education of new hosts regarding rules, norms and conventions of the non-professional hosting approach. Knowledge sharing and support takes place in host forums and groups that have been set up online and provide a valuable channel for communication and opinion exchange over host practices, roles and behaviours. Moreover, as hosts stated, they try to lobby to shape the regulatory landscape in Japan in order for it to become more favourable for the entire P2P-accommodation sector. Given the imposition of the recent regulatory framework, the sector and particularly individual hosts have become vulnerable in their ability to fairly compete against professional lodging businesses.

Inevitably, the role of P2P-accommodation host communities is imperative for the resilience of the entire sector in Japan but also the country's accommodation product offering.

Most importantly, the resilience of individual host communities is paramount for the sustainable development of the tourism industry of Japan. Through hosts' organized activities, tourism may become more socially inclusive and incorporate members of the local community (i.e. hosts) in the development and operation of the industry. Likewise, the behaviours of P2P-accommodation hosts contribute to a more beneficial context of operation, which may lead to the local community enjoying the advantages of P2P-accommodation exchanges. For instance, hosts protect local communities and neighbourhoods from Airbnb-related disruptions by promoting their interests and safeguarding against unruly guests' behaviours (Stergiou and Farmaki, 2019). In addition, as this chapter shows, hosts aim to restore the reputation of P2P accommodation in the eyes of the public by highlighting the positive aspects of P2P transactions. Hence, it may be argued that a form of collective individual responsibility (Kaufman, 2015) emerges through the establishment of host groups acting as a community of practice in the P2P-accommodation sector. Such actions are conducive to the promotion of more resilient, inclusive and sustainable communities.

Conclusion and Implications

As global tourism continues to grow, imposing potentially negative impacts on local communities (Bec *et al.*, 2016; Cheer *et al.*, 2019), the importance of community resilience in contributing to sustainable development is heightened. In fact, the rapid expansion of the P2P-accommodation sector in tourism destinations indicates that community resilience needs to be considered in the policy-making agenda and/or in efforts to regulate the sector. Within this context, the role of P2P-accommodation hosts as key members of the local community (Caldicott *et al.*, 2020) is elevated as their responsible behaviour might contribute to the strengthening of the resilience capacity of the local community (Farmaki *et al.*, 2019). We drew from the case of Japan, where P2P accommodation was growing steadily until the introduction of the Private Lodgings Business Act, to examine the perceived responsibilities of Airbnb hosts in order to illuminate understanding of how individual, non-professional hosts constitute a source of community resilience.

It appears that the new regulation has forced individual, non-professional hosts out of the Airbnb platform, leading to the rise of two issues. First, the number of professional hosts is increasing as individual, non-professional hosts are marginalized from the platform. This, in turn, threatens the resilience capacity of the local community. Hosts represent a key component of the host community (Caldicott *et al.*, 2020). Second, the professionalization of hosting in Japan means that the nature of P2P accommodation is changing as it becomes more profit centric, limiting the potential of visitor–host meaningful contact, which is key for community resilience. In response to such changes, Airbnb non-professional hosts seem to have entered into a race to restore the reputation and the public's understanding of P2P accommodation through activities such as lobbying, self-organizing and knowledge sharing. Indeed, according to Berkes and Ross (2013) community resilience is largely dependent on self-organization.

Given the absence of a formal and comprehensive training programme for hosts provided by the platform (Farmaki and Kaniadakis, 2020), host communities' self-organization tactics help to fill an important gap. Self-organized host communities can be useful in sharing guest experiences and in helping to promote a 'purist', non-professional hospitality style driven by non-financial motives such as international cultural exchanges and social experiences (Moon *et al.*, 2019). Thereby, through Airbnb non-professional host groups, the resilience capacity of the local community amid the continuous growth of global tourism may be strengthened. These host communities assume responsibility for their practices while attempting to formalize their role in the P2P-accommodation sector through lobbying activities. Simultaneously, hosts acknowledge the role of the platform in community resilience by highlighting the institutional responsibility (Helberger *et al.*, 2018) carried by platforms as imperative. As such, regulatory frameworks surrounding P2P-accommodation activity, need to consider not only hosts' and guests' responsibilities in policy making but also the role of platforms as they emerge as key influencers on users' behaviours and, subsequently, on local communities' prosperous development and well-being.

References

Airbnb (2018) Notification in accordance to the Private Lodging Business Act (minpaku). Available at: https://www.airbnb.com/help/article/2274/what-should-i-know-about-booking-a-home-in-japan (accessed 3 September 2020).

Bec, A., McLennan, C.L. and Moyle, B.D. (2016) Community resilience to long-term tourism decline and rejuvenation: a literature review and conceptual model. *Current Issues in Tourism* 19(5), 431–457.

Bec, A., Moyle, B. and Moyle, C.L. (2018) Resilient and sustainable communities. *Sustainability* 10(12), 4810.

Berkes, F. and Ross, H. (2013) Community resilience: toward an integrated approach. *Society & Natural Resources* 26(1), 5–20.

Braun, V., and Clarke, V. (2006) Using thematic analysis in psychology. *Qualitative Research in Psychology* 3(2), 77–101.

Caldicott, R.W., von der Heidt, T., Scherrer, P., Muschter, S. and Canosa, A. (2020) Airbnb – exploring triple bottom line impacts on community. *International Journal of Culture, Tourism and Hospitality Research* 14(2), 205–223.

Cheer, J.M., Milano, C. and Novelli, M. (2019) Tourism and community resilience in the Anthropocene: accentuating temporal overtourism. *Journal of Sustainable Tourism* 27(4), 554–572.

Cheng, M. and Jin, X. (2019) What do Airbnb users care about? An analysis of online review comments. *International Journal of Hospitality Management* 76, 58–70.

Cheng, M., Zhang, G. and Wong, A.I. (2020) Spanning across the boundary of Airbnb host community: a network perspective. *International Journal of Hospitality Management* 89: 102541.

Farmaki, A. (2015) Regional network governance and sustainable tourism. *Tourism Geographies* 17(3), 385–407.

Farmaki, A. (2019) Corporate social responsibility in hotels: a stakeholder approach. *International Journal of Contemporary Hospitality Management* 31(6), 2297–2320.

Farmaki, A. and Kaniadakis, A. (2020) Power dynamics in peer-to-peer accommodation: insights from Airbnb hosts. *International Journal of Hospitality Management* 89: 102571.

Farmaki, A. and Kladou, S. (2020) Why do Airbnb hosts discriminate? Examining the sources and manifestations of discrimination in host practice. *Journal of Hospitality and Tourism Management* 42, 181–189.

Farmaki, A., Stergiou, D. and Kaniadakis, A. (2019) Self-perceptions of Airbnb hosts' responsibility: a moral identity perspective. *Journal of Sustainable Tourism*, 1–21.

Fujita, J. (2019) Airbnb lobbies local Japan governments to ease curbs for Olympics. *Reuters*, 19 November. Available at: https://www.reuters.com/article/us-olympics-airbnb-japan/airbnb-lobbies-local-japan-governments-to-ease-curbs-for-olympics-idUSKBN1XT0WX (accessed 3 September 2020).

Goulding, C. (1999) Grounded theory: some reflections on paradigm, procedures and misconceptions. *Working Paper Series* WP006/99. University of Wolverhampton, Wolverhampton, UK.

Green, M.J. (2002) Institutional responsibility for global problems. *Philosophical Topics* 30(2), 79–95.

Hall, C.M. and Lew, A.A. (2009) *Understanding and Managing Tourism Impacts: an Integrated Approach*. Routledge, Abingdon, UK.

Helberger, N., Pierson, J. and Poell, T. (2018) Governing online platforms: from contested to cooperative responsibility. *The Information Society* 34(1), 1–14.

Hennink, M., Hutter, I. and Bailey, A. (2010) *Qualitative Research Methods*. Sage, London.

Kaufman, B. (2015) Theorising determinants of employee voice: an integrative model across disciplines and levels of analysis. *Human Resource Management Journal* 25(1), 19–40.

Leung, X.Y., Xue, L. and Wen, H. (2019) Framing the sharing economy: toward a sustainable ecosystem. *Tourism Management* 71, 44–53.

Lew, A.A. (2014) Scale, change and resilience in community tourism planning. *Tourism Geographies* 16(1), 14–22.

Linley, P.A. and Maltby, J. (2009) Personal responsibility. In: Lopez, S.J. (ed.) *The Encyclopedia of Positive Psychology*. Blackwell Publishing, Boston, Massachusetts, pp. 685–689.

Liu, X. and Chen, H. (2020) Sharing economy: promote its potential to sustainability by regulation. *Sustainability* 12(3), 919.

Magis, K. (2010) Community resilience: an indicator of social sustainability. *Society and Natural Resources* 23(5), 401–416.

McGregor, S.L.T. (2017) Consumer perceptions of responsibility. In: Emilien, G., Weitkunat, R. and Ludicke, F. (eds) *Consumer Perceptions of Product Risks and Benefits*. Springer, Cham, Switzerland, pp. 567–596.

Middlemiss, L. (2010) Reframing individual responsibility for sustainable consumption: lessons from environmental justice and ecological citizenship. *Environmental Values* 19(2), 147–167.

Midgett, C., Bendickson, J.S., Muldoon, J. and Solomon, S.J. (2018) The sharing economy and sustainability: a case for Airbnb. *Small Business Institute Journal* 13(2), 51–71.

Moon, H., Miao, L., Hanks, L. and Line, N.D. (2019) Peer-to-peer interactions: perspectives of Airbnb guests and hosts. *International Journal of Hospitality Management* 77, 405–414.

Nieuwland, S. and van Melik, R. (2018) Regulating Airbnb: how cities deal with perceived negative externalities of short-term rentals. *Current Issues in Tourism* 23(7), 811–825.

Ritchie, J., Lewis, J., Nicholls, C.M. and Ormston, R. (2013) *Qualitative Research Practice*. Sage, Los Angeles, California.

Robinson, O.C. (2014) Sampling in interview-based qualitative research: a theoretical and practical guide. *Qualitative Research in Psychology* 11(1), 25–41.

Selznick, P. (2002) *The Communitarian Persuasion*. Woodrow Wilson Center Press, Washington.

Stergiou, D.P. and Farmaki, A. (2019) Resident perceptions of the impacts of P2P accommodation: implications for neighbourhoods. *International Journal of Hospitality Management* 91: 102411. DOI: 10.1016/j.ijhm.2019.102411

Sthapit, E. (2019) My bad for wanting to try something unique: sources of value co-destruction in the Airbnb context. *Current Issues in Tourism* 22(20), 2462–2465.

Sthapit, E. and Björk, P. (2019) Sources of distrust: Airbnb guests' perspectives. *Tourism Management Perspectives* 31, 245–253.

Strauss, A. and Corbin, J. (1990) *Basics of Qualitative Research: Grounded Theory Procedures and Techniques*. Sage, Newbury Park, California.

Sugiura, E. (2019) Airbnb begins to recover in Japan, a year after crackdown. *Nikkei Asia*, 17 April. Available at: https://asia.nikkei.com/Spotlight/The-Big-Story/Airbnb-begins-to-recover-in-Japan-a-year-after-crackdown (accessed 3 September 2020).

Sundararajan, A. (2016) *The Sharing Economy: the End of Employment and the Rise of Crowd-based Capitalism*. MIT Press, Cambridge, Massachusetts.

The Mainichi (2018) 'Minpaku' lodging law goes into effect, but complicated process thwarts property owners. *The Mainichi*, 15 June. Available at: https://mainichi.jp/english/articles/20180615/p2a/00m/0na/028000c (accessed 3 September 2020).

Wang, Y., Asaad, Y. and Filieri, R. (2020) What makes hosts trust Airbnb? Antecedents of hosts' trust toward Airbnb and its impact on continuance intention. *Journal of Travel Research* 59(4), 686–703.

Wenger, E. (1998) *Communities of Practice: Learning, Meaning, and Identity*. Cambridge University Press, Cambridge.

4 Peer-to-peer Accommodation and Resilient Hosts in Split: the Case of Radunica Street

Helena Tolić* **and Ingeborg Matečić**
University of Zagreb, Croatia

Abstract
Through a qualitative approach, including autoethnography and open-ended interviews, this study focuses on how offline peer-to-peer (P2P)-accommodation trading was used as a post-crisis recovery tool. We focused our study on Radunica Street, situated in the city centre of the Croatian city of Split. We identified the motives that led hosts to engage in offline P2P-accommodation trading and then, eventually, online P2P-accommodation trading and uncovered resilience attributes that characterize these individuals. Community networks and new relationships caused by offline P2P-accommodation trading and the cooperation between hosts and **gatherers** (locals who searched for tourists) were analysed. Our research explains how this arbitrary and unstable cooperation within the local community shaped its people. Moreover, it led us to a broader definition of the term 'peer-to-peer accommodation' since the results show that on Radunica Street this does not apply only to a 'host-to-tourist' way of doing business. Before paid-online P2P-network platforms entered the Croatian market, this 'peer-to-peer' relationship was actually based on a triangular 'gatherer–host–tourist' cooperation, which has changed the culture of everyday life in the street for good.

Keywords: resilience, peer-to-peer accommodation, local community, hosts, accommodation trading

Highlights

- The local community of Radunica Street developed community resilience and responded to various post-war crises using offline peer-to-peer (P2P)-accommodation trading.
- The motives for starting offline P2P-accommodation trading are singled out.
- The attributes of resilient hosts are identified.
- The community network and relationships are described under the new circumstances caused by offline P2P-accommodation trading.

Introduction

From a broader social perspective, community resilience is an ambiguous concept described differently by scholars in various research disciplines while no common agreement exists on the definition of the term (Patel *et al.*, 2017). Nevertheless, Patel *et al.* (2017) identified nine elements shared across numerous definitions of community resilience related to disasters. These 'core elements' (Patel *et al.*, 2017, p. 1) are: (i) local knowledge; (ii) community networks and

*helena0908@gmail.com

(eds Anna Farmaki, Dimitri Ioannides and Stella Kladou)
DOI: 10.1079/9781789246605.0004

relationships; (iii) communication; (iv) health; (v) governance and leadership; (vi) resources; (vii) economic investment; (viii) preparedness; and (ix) mental outlook.

The research on resilience in tourism generally investigates the post-system collapse of highly vulnerable systems caused by a disaster (Coaffee and Wood, 2006; Larsen *et al.*, 2011) but also how communities react to changes induced by tourism (Farrell and Twining-Ward, 2004). Despite various definitions of community resilience, this research relies on the following two definitions. The first describes community resilience as 'the existence, development and engagement of community resources to thrive in an environment characterized by change, uncertainty and unpredictability' (Magis, 2010; Bec *et al.*, 2016, p. 432) while the second states that it is 'the ability of communities or groups to adapt or cope with external stress and disturbances, caused by social, political and environmental change' (Adger, 2000; Bec *et al.*, 2016, p. 432). In other words, social but also economic resilience reflects the ability to absorb and adapt to change to ensure ongoing economic and system functioning (McKercher and Prideaux, 2020). In our specific case, we define community resilience as being achieved through the use of individual and community resources such as inhabitants' own homes, apartments or single rooms, and even public infrastructure to conform and adapt to external disturbances caused by social, political and economic change.

Both resilience and peer-to-peer (P2P) accommodation have, in recent years, attracted much scholarly attention. In relation to the latter term, many scholars have, since 2010, focused mostly on Airbnb as the most prominent online P2P-accommodation trading platform (Reinhold and Dolnicar, 2018b, c, d; Dolnicar, 2019; Kuhzady *et al.*, 2020). It is well worth mentioning that the *Annals of Tourism Research* launched a curated collection in 2019 where past as well as the most recent academic insights on the phenomenon of paid, unpaid, online P2P accommodation and its networks can be found (Dolnicar, 2019). Nevertheless, research on offline P2P-accommodation trading and community resilience is scarce. The impacts of short-term rentals and possible approaches to manage these has been researched from the residents' perspective (Park and Agrusa, 2020; Stergiou and Farmaki, 2020), and the former study even provided a resident typology according to perceptions of P2P-accommodation impacts on neighbourhoods. However, both studies restricted their focus to online P2P-accommodation networks.

Our aim is to showcase how the residents of Radunica Street in the Croatian coastal city of Split (henceforth called Radunica) developed community resilience while responding to various social and financial crises following the Croatian War of Independence, first through, what we term, offline P2P trading of their own rooms and apartments and subsequently through paid-online P2P-accommodation trading. In our case, the term paid-offline P2P-accommodation trading is associated with both informal accommodation trading as well as illegal lodging and it refers to short-term rentals. The reason we use the term paid-'offline' P2P accommodation is that the economic activity of P2P accommodation within the Radunica community started long before online platforms entered the Croatian tourism market, and it was connected to an illegal economic activity but still as a form of 'peer-to-peer' activity.

Our research results flesh out the motives driving these individuals to embrace offline P2P-accommodation trading while recognizing host attributes, which can be described as resilient behaviour. Additionally, we were able to reveal the residents' emotions that were stirred through the P2P-accommodation phenomenon while trading their own homes, as well as the community networks and relationships occurring under the new economic and trading circumstances on this particular street.

Theoretical Background

P2P accommodation

Paid P2P accommodation is often associated with the notions of P2P-accommodation network platforms, P2P trading, the sharing economy, or collaborative consumption. Terms like the sharing economy or collaborative consumption are frequently used to describe P2P trading, although neither sharing nor collaboration best describes the notion of paid P2P accommodation (Dolnicar, 2019). Several researchers have analysed and debated the concept of sharing in the context of P2P accommodation (Belk, 2014; Habibi *et al.*, 2016; Breidbach and Brodie, 2017; Reinhold and Dolnicar, 2018a; Dolnicar, 2019). Generally, they agree that sharing does not fully apply to the term paid P2P accommodation. Since sharing implies, among others, 'use[ing], occupy[ing], or enjoy[ing] (something) jointly with another or others' (*Oxford English Dictionary*, 2020a) this certainly is not the case with the contemporary P2P-accommodation concept, which normally applies to P2P accommodation available for trading on Internet platforms. Furthermore, as Dolnicar (2019) argues, the term collaborative consumption does not accurately define paid P2P accommodation either because the verb to collaborate means '[to] work jointly on an activity or project' (*Oxford English Dictionary*, 2020b), which is not what the hosts who rent their apartments do with their guests.

Dolnicar (2019) defines paid-online P2P accommodation as 'space suitable for overnight stays sold by a non-commercial provider (the host) to an end user (the guest) for short-term use through direct interaction between host and guest' (2019, p. 248). According to Dolnicar (2019) there are several dimensions of the definition, besides paid and online attributes, which are worth pointing out to better construe the subject of our case study. These dimensions, or as Dolnicar calls them 'key building blocks' (2019, p. 248) include:

1. *Space* – includes different kinds of spaces suitable for overnight accommodation, ranging from low-quality shared rooms to luxury residences, targeting different market segments (Lutz and Newlands, 2018; Dolnicar, 2019).
2. *Non-commercial provider (the host)* – the host's primary aim is often associated with gaining additional income for purposes like paying off a loan, covering basic expenses or purchasing a new appliance or just for socializing and meeting interesting people (Ikkala and Lampinen, 2015; Karlsson and Dolnicar, 2016; Hardy and Dolnicar, 2018). Conversely, commercial accommodation providers aim to boost profits and maximize return on investment (Dolnicar, 2019).
3. *Direct interaction* – if the interaction between host and guests is direct rather than through an intermediary then trading occurs between two peers (Dolnicar, 2019).

We conclude that paid-online P2P accommodation is a space traded on online P2P-accommodation network platforms. This definition becomes more complex considering that these networks trade other types of accommodation and serve as distribution channels for professional tourism accommodation businesses (Dolnicar, 2019). Out of Booking.com's total of 27 million listings in 2018, which included professional tourism accommodation facilities such as hotels, resorts and hostels, 5 million represented private homes, apartments, homestays and other unique lodgings (World Bank Group, 2018, p. 15). Meanwhile Airbnb, which began as an online P2P platform trading private accommodation, opened the platform to hotel distribution (May, 2018).

At the micro level, some studies are concerned with the changing nature of homes (Stabrowski, 2017; Roelofsen, 2018) and how paid-online P2P accommodation alters them into 'spaces of domestic entrepreneurialism' (Stabrowski, 2017, p. 327) changing the way hosts use and decorate them. At the macro level and within residential areas, impacts of paid-online P2P accommodation on the everyday life of residents range from increased home prices (Barron *et al.*, 2020) to increased rents (Schäfer and Braun, 2016; Gurran and Phibbs, 2017; Horn and Merante, 2017; Barron *et al.*, 2020).

Previous research on the motives behind the renting of private accommodation shows that the main reasons that hosts choose to do this fall within three categories: (i) income; (ii) social interaction; and (iii) sharing (Karlsson and Dolnicar, 2016). Income has further subcategories 'to pay the bills', 'for money' or 'to afford luxury'; social interaction is about 'meeting people', 'for the love of it' or to overcome loneliness (Karlsson and Dolnicar, 2016; Farmaki and Stergiou, 2019); and the motives for sharing are reflected in subcategories such as to make use of 'unused space', 'sharing my world' and 'sharing my house' (Karlsson and Dolnicar, 2016).

Community resilience in tourism

A community can face rapid changes caused by disasters such as earthquakes, tsunamis or terrorist attacks but major human-scale changes occur on a constant and continuous basis leading to, for example, biodiversity losses or the destruction of certain cultural traditions (Lew, 2018). Slow drivers of change range from the effects of various economic crises to climate change. These influence resilience (Lew, 2018). As Bec *et al.* (2016) point out resilience frameworks in tourism fail to explain community resilience to long-term structural economic change driven by tourism. They conclude that 'further research is needed to explore how community resilience can be utilized as a tool to respond to and manage long-term structural change' (Bec *et al.*, 2016, p. 432). Nevertheless, the literature recognizes dimensions of community resilience to long-term structural change, such as: (i) capacity building, collaboration, experience, communication and knowledge sharing (Paton and Johnson, 2001; Zautra *et al.*, 2008; Berkes, 2009); (ii) partnerships/networks (Holladay and Powell, 2013); (iii) support (Norris *et al.*, 2008); (iv) adaptability (Folke *et al.*, 2005); and (v) identifying opportunities/positive outlook (Folke *et al.*, 2005). A community's capacity to respond to structural change is connected to its competence to adjust to disturbances, diminish hardship, take advantage of opportunities, and handle the transformation that arises (Gallopín, 2006; Linnenluecke *et al.*, 2012; Bec *et al.*, 2016).

Lew (2014) has proposed a scale, change and resilience (SCR) model, establishing four types of tourism contexts, based on the degree of disturbance (from gradual change to unexpected disturbance) and the scale of tourism stakeholders (from private entrepreneurs to public sector such as local or regional governments). Essentially, the SCR model describes 'specific resilience issues, methodologies and measurements, ranging from entrepreneurs managing daily maintenance needs, to community disaster readiness, response, and recovery' (Lew, 2014, p. 14).

Community resilience in tourism has been researched from the perspective of overtourism and social-ecological resilience (Cheer *et al.*, 2019), rural tourism and sustainable tourism development (Ahmad Fitri Amir *et al.*, 2015), perceptions of tourism impacts and community resilience to natural disasters (Tsai *et al.*, 2016), community sustainability and resilience (Lew *et al.*, 2016) and so on. Meanwhile, research on community resilience relating to short-term rentals or offline P2P accommodation trading is scarce. In the next section we provide details of a case study of offline P2P-accommodation trading in one particular street in the city of Split and showcase aspects of the local community's resilient behaviour.

Case Study: Offline P2P-Accommodation Trading in Radunica, Split, Croatia

Split, a city located on the Dalmatian shore of the Adriatic Sea, is one of Croatia's most important tourism destinations. Before the beginning of the 21st century, Split was mostly known as a 'transit city' for tourists. The reason for this lies in its geographical position and good transportation links to islands and other, more attractive, holiday destinations. Although Split was a tourism city, it was not as popular as other Croatian tourism destinations like Dubrovnik or Opatija. Therefore, its tourism infrastructure was not as developed, as in the other destinations, especially with regards to private accommodation.

Even though Split had a long history of tourism development, it is only during the first decade of the 21st century that it emerged as one of Croatia's top tourism destinations. Demand for tourist accommodation in the city grew rapidly, indicated by the increase of tourist arrivals and overnight stays (Table 4.1).

As in many Dalmatian towns, the city centre of Split and its closest, not to mention oldest, neighbourhoods form a charming net of *kaletas*. *Kaleta* is a word in the Dalmatian dialect for an old, narrow and small street with specific Mediterranean vernacular architecture. One of those streets is Radunica, the focal area of our research.

Table 4.1. Total number of tourist arrivals and overnight stays in Split from 2012 to 2019. From Croatian Bureau of Statistics, 2020a.

Year	Domestic and foreign tourist arrivals	Domestic and foreign overnight stays
2012	265,630	678,401
2013	318,057	861,606
2014	381,227	1,045,727
2015	487,474	1,339,598
2016	583,041	1,717,396
2017	720,325	2,127,350
2018	859,224	2,494,072
2019	941,185	2,734,632

Radunica is the central street of Lučac, one of the oldest neighbourhoods in the town outside the historical city centre. It is located approximately 5 minutes on foot east of the famous Diocletian's palace. In the past, Radunica was inhabited by peasants, mostly farmers and fishermen, and it is from them that it inherited its specific architectural features – modest stone houses with gardens (Piplović, 2006, p. 454).

Just like other *kaletas* in the historical city centre, Radunica was among the first streets in Split whose local inhabitants were involved in offline P2P-accommodation trading activity. Before 2004, this activity rarely occurred on this street although the concept of short-term rentals of people's properties to tourists had been present in Split and the rest of the country even before the break-up of the former Yugoslavia. During the summer of 2004, a couple of locals on Radunica readapted their apartments for tourists' use. Since then, every summer, more local people have been doing the same, and today, along Radunica, almost every stone house has a sign indicating the availability of apartments for rent.

Based on the results of the first part of our research (2004–2014), we addressed the question: How was offline P2P-accommodation activity organized along Radunica before the advent of global Internet trading platforms, like Booking.com and Airbnb.com, in Croatia? Tolić (2019) describes how, starting from 2004, the number of offline P2P-accommodation family businesses on Radunica began growing rapidly each year. This economic activity, which was unreported and hence illegal, was organized and controlled by locals who created a cooperation network of **hosts** and **gatherers**. Hosts were those locals who rented their apartments or rooms to tourists, while gatherers were the locals who searched for tourists (usually at the bus stations) and brought them to stay in the host-owned accommodation.

As to *why* so many locals on Radunica strongly embrace paid-offline and, subsequently, -online P2P-accommodation trading we base our analysis on resilience theory. We believe the community's involvement in offline P2P-accommodation trading reflected a response to the prevailing socio-economic context in Croatia at that time. This was a period when economic instability was increasingly evident, as reflected in the high unemployment rates throughout the country. Split and Radunica were particularly severely affected by this situation (Table 4.2), leading residents to seek alternative ways of income generation, such as through renting their properties as holiday accommodation.

This financial crisis was prolonged because, in addition to various inherited national problems following the Croatian War of Independence (1991–1995), including widespread economic restructuring and unsuccessful privatization processes, the global economic collapse of 2007–2008 worsened matters. This period was especially traumatizing, and a great number of locals of Radunica were forced to find ways to survive. For many, the solution was to perceive their homes and the street itself as a tourist resource.

To overcome economic hardship, these individuals did more than simply remodel their homes or adapt their space for the tourists. Their homes were transformed from spaces of social reproduction into places of (paid) work to be shared with tourists. This initiated deep-set structural

Table 4.2. Unemployment rate in Croatia and Split-Dalmatia County in the 21st century.[a] From Croatian Bureau of Statistics, 2020b.

	Unemployment rate	
Year	Croatia (%)	Split-Dalmatia County (%)
2001	22.6	30.1
2002	23.6	30.8
2003	20.6	27.4
2004	18.9	24.2
2005	19.0	23.7
2006	17.7	22.2
2007	16.3	21.4
2008	14.3	19.1
2009	15.1	19.4
2010	18.2	22.7
2011	19.1	24.0
2012	19.7	25.8
2013	21.5	27.9
2014	22.2	28.6
2015	19.3	26.1
2016	16.9	24.1
2017	13.9	21.4
2018	11.1	18.2
2019	9.1	15.4

[a]The unemployment rate is expressed as a percentage of the labour force (total number of employed and unemployed).

changes in the community's everyday life, culture and traditions since it altered traditional social relationships from **just neighbours** into **business-associates-and-competitors**. Thus, we have focused on: (i) the attributes of the resilient members of the local community who responded to the financial crisis; (ii) the community networks and relationships that supported the resilience of the local community; and (iii) the significance of paid-offline P2P-accommodation trading for the community's resilience.

Methodology

One of the authors (Tolić), who grew up on Radunica, witnessed the street's transformation over the years. Since her family was one of the first families on Radunica to embrace P2P-accommodation trading as hosts, through the years Tolić participated in all activities related to the business. During that time, she embarked on the longitudinal study that ran until 2018. From 2004 until 2014 the offline P2P-accommodation trading on Radunica was at its peak before most hosts shifted to online rental platforms. We analysed this period through a qualitative approach using open-ended interviews, autoethnography and phenomenology.

Phase two of the research occurred during the summer of 2020, when we conducted 13 interviews with members of the local community. Five interviewees were hosts, five were gatherers who eventually became hosts, one was a cleaning lady, one managed Booking.com accounts for many hosts, and one had been both a host and a gatherer since the 1960s. Saturation in this qualitative research was reached and 'no new information or themes are observed in the data' (Guest *et al.*, 2006, p. 59; 2020). Eleven interviews were held in person, and two were conducted by phone. The questions were prepared in advance as guidance, but new questions were added or changed on site when an interviewee steered the discussion in a new direction.

Our original plan was to also research the subject autoethnographically in 2020, but the COVID-19 crisis stopped us from pursuing that goal. In 2020, the tourist numbers were extremely low on Radunica making it impossible to describe the subject through our own lived experience and to analyse sociocultural processes connected to P2P-accommodation trading. However, since this research follows on from the earlier study of offline P2P accommodation on Radunica, we were still able to use autoethnography for the purpose of setting the scene.

Results

The beginnings of offline P2P-accommodation trading on Radunica and resilient behaviour

Before the early 2000s, the concept of offline P2P-accommodation trading was rare but still present on Radunica. During the late 1960s, only a couple of families in the whole Lučac area rented out rooms in their own apartments directly to tourists. Dana, an 85-year-old local was among the few who did:

> On Christmas day of 1969, I took in my first guest. I was one of the few here, if not the only one to do something like that. My husband was a **carrier** [carried tourist's suitcases for money], so we decided to do it.
>
> (Dana, bold added)

As in Dana's case, until the early 2000s, informal accommodation for visitors in Croatian tourism destinations was provided through so-called **household rentals**, meaning a tourist was a paying guest in the family's home while the family itself was also living there. On Radunica, only a couple of working-class families with low family income engaged in household rentals, and these families were mostly newcomers from the surrounding villages. The accommodation was modest, lacking any type of luxury. The whole family would sleep in the kitchen, or in one bedroom (depending on their apartment's size), while the guests had their own room in the same apartment and shared the bathroom. As Roko, one of the locals tells us: 'Household rentals existed so that on that day, you wouldn't have to eat only *chicken bologna*, but that you could afford ham.'

Indeed, most locals regarded it as an economic activity for poor people who could not make ends meet. As Marin, whose family rented out their home during the 1970s and 1980s, explains:

> Other people were embarrassed to do it, they worked for a pay cheque. ... The need, I guess, the financial need probably and poor [financial] conditions [were the motive]. Surely no one started renting at that time out of wealth, but out of deficit. Understand me, eh? ... In the beginning, it was only the poor, and some of them built castles out of it [became wealthy], you know ...
>
> (Marin)

Thus, a label of shame was attached to people who rented their homes out before the early 2000s. In those days, today's notions that renting out one's property breeds entrepreneurship or allows hosts to meet people from other cultures were non-existent. The activity was perceived as a situation that compromised a family's privacy or a home's hygiene by letting foreigners sleep in your own bed and share your bathroom, not to mention the danger of letting strangers into your house. There was also a negative feeling towards the bag carriers who would transport baggage, and sometimes the tourists themselves, on a trolley for money. As both Roko and Marin hinted, in most cases this belief of extreme poverty of carriers and hosts during the 20th century was, in fact, mostly false:

> Within 2 or 3 months you could buy an apartment ... Normally, everyone thought – they are poor – you know, they would look at those people with contempt, that's how it was. If only I was smarter, but I was embarrassed, I would have done it [be a carrier] for 5 years and I wouldn't have to work anymore a single day in my life.
>
> (Marin)

Dana, whose husband was a carrier and who rented rooms in her own apartment confirms:

> My husband earned an apartment for my daughter, a car for my son, a garage, a hotel in the centre of the [Diocletian's] palace with 12 beds – everything with his work. ... And while they [other neighbours] were all going to church, praying, I was working. Now that I am old and can't work no more, now I can go to church, and I go.
>
> (Dana)

This last remark in Dana's statement captures the resilience of local people involved in paid P2P-accommodation activity on Radunica during the 20th century. They were faced with financial troubles, and years of tacit social condemnation. The statements presented show the attributes that helped the local community of Radunica face those challenges:

- *Willingness to work* – to change an unsatisfying situation.
- *Indifference* – they reflected indifference towards the predominant opinion within the local community.
- *Contrarianism* – they did not stop although it was against the local cultural norm of that time.
- *Bravery* – they worked despite condemnation, labels of shame and the fact it was an illegal business.
- *Persistence* – to continue being involved with P2P-accommodation activity throughout decades and conform to new market rules.

P2P-accommodation trading on Radunica and resilient behaviour – the middleman

Around 2004, offline P2P accommodation in the local community started gaining momentum. It quickly emerged as the predominant occupation of many locals and began to change their everyday life. The role of the middleman was extremely important. First, during the offline era of P2P-accommodation trading, the middlemen were local gatherers, while after the online platforms entered the market, this role was assumed by individuals who ran the business on the Internet for the hosts.

From 2004, local inhabitants started organizing an illegal and tightly cooperative network of gatherers and hosts: this lasted for 9 years (Tolić, 2019). Tolić (2019) describes how the gatherers either lacked spare accommodation or their property was too small to be let out to tourists. This led them to cooperate with their neighbours who were hosts by finding tourists (clients) for a commission. Meanwhile, the hosts were the people who had spare space but lacked access to tourists; they were either unwilling to search for tourists or too busy working at their regular jobs. The gatherers would go (mostly) to the main bus station where they would wait for the arriving tourists, very often working 12–16 hours/day during the high season. They offered the accommodation on Radunica to tourists at the bus station while negotiating and setting the price. Then, the gatherers would escort the tourists from the bus station to the apartments, where the hosts would pay them the commission. Thus, this offline P2P-accommodation business was not a two-sided cooperation between the host and the tourist, but a triangular P2P cooperation between a host, a gatherer (middleman or intermediary) and a tourist.

> I was bringing the tourists all around the neighbourhood, they [hosts] would just call me ... [We worked] mostly illegally, yes. Not everyone, but the vast majority [worked illegally], me first. So what?! [You work illegally] a little during the summer, what will happen to you? Then they started with inspections, they started with the Internet, everyone started to be afraid of it and it [the need for a gatherer] stopped ...
>
> (Petra)

When the government began regulating the market through inspections and fines, more and more hosts began working legally. Even under these circumstances, the cooperation between hosts and gatherers continued for a few more years after the online network platforms such as Booking.com entered the Croatian tourism market. Online platforms marked the beginning of the end of this kind of cooperation between hosts and gatherers on Radunica. Certain local gatherers found new ways to resist financial instability by becoming hosts. Others simply stopped working. Eventually, the need for gatherers' services disappeared.

With the rise of Booking.com, and then Airbnb.com, hosts were able to upload their apartments online, often with the help of their children. Many gatherers could not keep up with these changes, as Roko explains:

> So that business [of a gatherer] started falling apart a few years ago. ... As long as they were in the Balkan-state-of-mind – 'I don't know English, what will I tell him, how will he pay me' – for me, for her [Petra] and for about 50 of us in the city, this was a Mecca. But unfortunately, times have changed, and hosts have children [who do these jobs for them now].
>
> (Roko)

For those who were unable, for several reasons, to advertise their own apartments on Booking.com, a new type of middleman emerged. These were people from all over Split, not necessarily from Radunica, who assumed the role of gatherer and did the job online instead of the hosts. They uploaded photos and added descriptions and communicated with the tourists. In other words, they replaced the host in online dealings/communication with the tourists. Although Booking.com is primarily a platform listing hotels and official tourism accommodations, the new middlemen found a way, a grey area of the business, to circumvent the company's rules. Effectively, they represented multiple hosts so they could *act* like a hotel. Maria, who takes bookings on Booking.com for many, states:

> There was a link for Booking in 2011 which I dug up to apply my cottage on the island ... I wanted to put an apartment there but there was no chance ... so I've put up my friends [friends' apartments] from Split. ... there had to be a minimum number of apartments, six ... and all at one address.
>
> (Maria)

Like the gatherers before, Maria does not report her business activity to the authorities. She only has a verbal agreement with hosts who pay taxes and fees to Booking.com, and a commission to her. After losing her job, she found the way to operate in the grey area of the business – as the middleman she works illegally with legal Internet platforms for the hosts who own legal businesses.

> I went to the University of ... but I also have a 'degree of the ghetto university'. ... You have situations when you must meet a tattooed, shaved guy and count €20,000–30,000 at the street corner. I did it all. At first, with Booking [Booking.com], money wasn't transferred over an account. These little fingers counted it [money] all, ... and there was blackmailing ...
>
> (Maria)

The role of middlemen since 2004 was crucial for developing P2P accommodation on Radunica. Individuals who became gatherers had no resources of their own (an apartment to rent) and basically *made up* a new job for themselves, which helped the community to prosper. Online P2P-accommodation trading made it possible for a new *group* of individuals to become a part of the process. Unlike the previous gatherers, they spoke foreign languages and were digitally literate. Interviews confirm the fact that both types of middlemen worked illegally and had similar attributes of creativity and did not hesitate to bypass the rules.

Acceptance of new circumstances and motivation for starting a P2P-accommodation business

The locals' motives to rent out their properties to tourists in the early 2000s reflect contingencies they experienced because of the national socio-economic context. By analysing interviews with hosts and gatherers, we isolated the motives guiding the involvement of the local community in paid-offline P2P accommodation (Fig. 4.1).

Due to Croatia's unsatisfactory economic situation and the global 2008–2009 financial crisis most of the motives were connected to financial difficulties and unemployment. Specifically, respondents highlighted the loss or instability of family income and unemployment as important motives. Indeed, only two reasons were unrelated to finances or the respondents' employment situation: (i) the wish to meet people from different cultures; and (ii) the desire to abandon a job with night shifts.

Mina, a 60-year-old woman, describes her motives:

> Well [my pension is] 1450 kunas [€192]. My husband has 2600 [€344] ... My son is not working. My daughter was in college, ... so then we went for it. ... You see someone working with tourists, and you have that one room [available, and you think] 'why shouldn't I when he can' and that's it.
>
> (Mina)

Although offline P2P accommodation seemed to be an obvious solution to solve their problems, choosing to embrace this activity was not easy nor was everyone comfortable with this solution. The lack of choice was hard for some, especially for the baby boomer generation who had to start all over, compete with the younger generation, and work harder in their own homes, despite their expectations of a peaceful retirement.

> A lot of people would stop doing this, at least they say so, they would give up if they could live normally. This is not extra [money], for us, this is not extra income for the majority that I know. It is not additional ... you live from this. What would we do without it? I don't know. We would kill each other. Everybody leaves [Croatia] anyway, we would go away, everyone leaves.
>
> (Petra)

While early on it was a long way from a financial crisis, the lack of other business opportunities

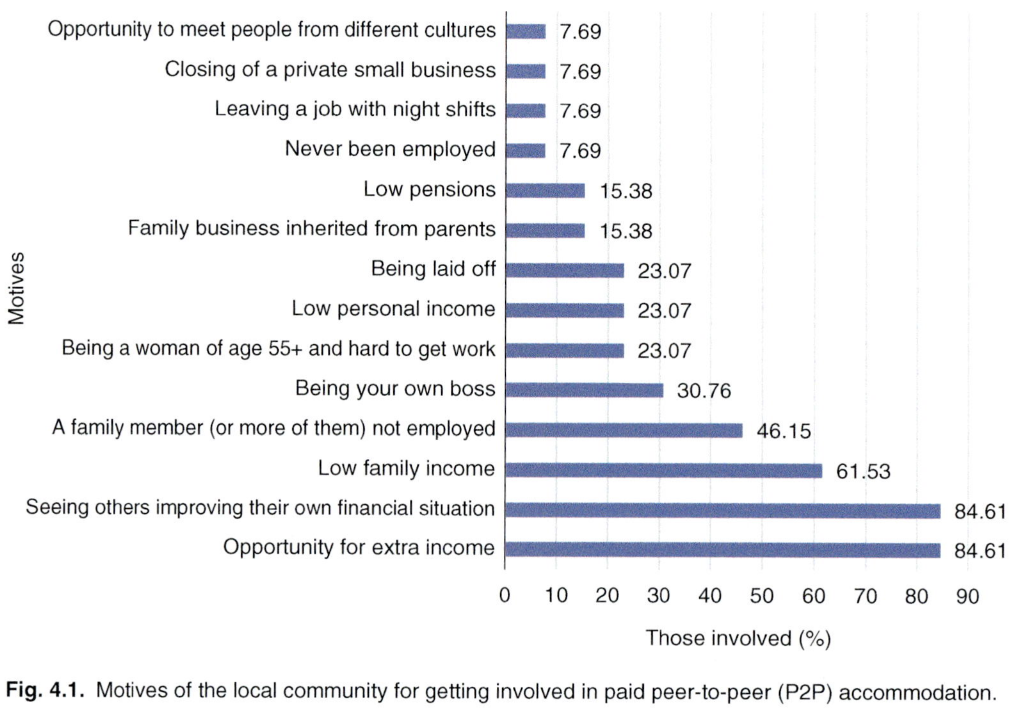

Fig. 4.1. Motives of the local community for getting involved in paid peer-to-peer (P2P) accommodation.

pushed many locals to become financially dependent on P2P accommodation. Nevertheless, this allowed a series of negative factors to arise in the community. These include: (i) distrust of and disappointment in power holders; (ii) feeling of underappreciation among baby boomers; (iii) loss of conventional feelings about one's home (redefining one's own home as a place of work); (iv) feelings of shame; (v) existential fears; (vi) anxiety of being caught working illegally; (vii) business competitiveness among locals who were once friends; (viii) loss of the feeling of belonging to the local community; (ix) distrust in the honesty of neighbours; (x) street depopulation; and (xi) loss of local traditions shared by the community.

To accept the changes offline/online P2P-accommodation trading brought to everyday life on Radunica, locals demonstrated resilience by acting, compromising and accepting the novel situation and new circumstances it entailed, regardless of whether they perceived them as positive or negative. In the following section, we show the attributes of those individuals who successfully navigated through these new circumstances.

Attributes of resilient individuals – affirmative outlook and turning a blind eye

In preparing for the field research, we created a series of research questions based on the previous investigation and personal experience that Tolić gained from her involvement in paid P2P accommodation activity on Radunica. She took notes, capturing her own thoughts on paid P2P accommodation on the street and changes this activity provoked in the community. Some were positive, like the financial effects or the restoration of the landscape. Others were negative, like the loss of traditions and togetherness, the exodus of locals from Radunica and gentrification. It became evident that the resilient individuals who started renting in 2004 and are still involved in the business have significantly improved their financial situation. These individuals have found it easier to get past nostalgia and feelings of loss. They showed no regrets regarding the negative influences of P2P accommodation while accepting that changes were a part of *something to be expected*:

> I feel sorry that [people are leaving], I mean, actually, in fact I am not sorry. It's fine by me. It is like that everywhere; it's how it goes. ... Well, now the suburbs are populated. The city is abandoned, the schools are empty. ... Apartment prices used to be lower, now they are high. But Split is being built, it progresses, this is normal because Split is now a tourist city.
>
> (Nina)

> It became a madhouse [since tourism developed in Split]. Everyone wants more, I want two beds more, he wants two tables more, a third one wants to print on pyjamas in the centre of the city promenade [laughs] ...
>
> (Roko)

Although they are aware of several negative effects that P2P-accommodation trading has inflicted on the city and on their own lives, the residents choose to focus on just the positive aspects. Paid P2P-accommodation activity has given them the opportunity to overcome the challenge of financial instability, so they have accepted the good together with the bad. Moreover, they downplay the importance of factors which influenced them negatively. Their narratives mostly come down to the decision *not to care* and *to turn a blind eye* towards the negatives because to do otherwise would undermine their primary focus – financial stability. Petra talked about local people moving out of the city centre permanently due to rapid tourism development in the city, and stated her own view of the situation:

> People hang out less, they are constantly in a hurry. ... There is not a living soul here from January till April. ... I don't know if anybody still lives there, they said about 500 [people] still live in the centre, it's all being rented out now. And everyone closes their stores, they open in April and close in October. ... The shops, restaurants, cafés, fast food, bakeries, all empty, it's all for them [tourists]. I don't care about it at all, and what they say about the charm [of the city centre disappearing], that doesn't touch me at all. I'm doing my own thing.
>
> (Petra)

The results made us realize that all the respondents had negative experiences on a personal level in the first years of becoming hosts or gatherers. The latter group were especially negatively affected. These experiences occurred because, among various challenges, they were chased by inspectors and fined, cheated out of money, or because other gatherers 'stole their guests'.

> Of course, there was [conflict], there were junkies there. ... like the late Bruno. You must buy him a beer, so he is nice to you ... He would go crazy on the old ladies [older gatherers], though. What can you do [but to] get away from him ... until he finds his guests and then he buys himself a beer, gets drunk and he's gone.
>
> (Petra)

Adaptability and cunning were important characteristics of the gatherers. Many mentioned facing shame, insecurities, fears and doubts in the first years of their work. They were exposed to aggressive behaviour at the bus station, including verbal and physical attacks. The ones who remained in business were those guided by internal motives:

> In the beginning, I was ashamed [to be a gatherer] ... And then everyone made fun of me ... I did it despite everyone ... Just about everyone told me that it's crazy, both my mother-in-law and my own mother. ... That resentment [motivated me], [and I wanted to] prove to them that I can do it.
>
> (Nina)

Resilient individuals do not become emotionally attached to occurrences that could endanger their goal. In the interviews they describe their own defiance, refutation, disregard or indifference towards the negative influences of P2P accommodation. What they all showed is the passion and the grit, the willingness to succeed by their own standards. They can retain an affirmative outlook and fixate on it. However, besides their personal characteristics, these individuals had to lean on the community and the networks of support they created to develop offline P2P accommodation on Radunica before online network platforms entered the market.

Resilience, community networks and relationships – culture of conflict vs culture of cooperation

The offline P2P-accommodation trading on Radunica was mostly illegal. It was organized spontaneously by locals who played the roles of either hosts or gatherers. These roles significantly changed the social relationships of people who were once just neighbours, who became business partners or business competitors, while bringing a new order on the street and creating a culture of cooperation but also of conflict (Tolić, 2019). These two variations coexisted in the community. The gatherers engaged in verbal business deals with the hosts. Thus, businesswise, the neighbourhood was divided between several gatherers. The deals they made were not obligatory, and some of the hosts worked with several gatherers. Often, this provoked distrust and disillusion between people on Radunica, meaning some stopped being friendly or talking to each other. However, the culture of conflict was significantly more evident between the gatherers at the main city bus station. It was based on competitiveness in an improvised market:

> You know how it is on the street, there were funny things and then there were fights, beating, everything. Because it's a street, anything can happen. ... Do you remember Anka [one of the gatherers]? Poor thing, she was good, but crazy. You need to get away from such people, you need to run away. ... When someone would steal my guests away from me, I would just go away, I wouldn't argue with anybody. ... Others would argue, they would fight, it was a madhouse [between the gatherers at the bus station].
>
> (Nina)

Conflicts were a destabilizing factor in the community and led to divisions between people, but they also strengthened relationships between most locals on Radunica. Although aware of the conflicts, they saw them as insignificant fights, largely occurring at the bus station, outside the community. Indeed, the gatherers who participated in conflicts were slowly marginalized by the community. During every new summer, fewer people would cooperate with these troublemakers. The ones who became most successful, in terms of developing a dense cooperation network, built their relations on trust, support, encouragement and integrity:

> There were four of us [gatherers who were sticking together], and until all four of us found guests, no one would go home. They would do it for me, and I would do it for them. If one of us couldn't find a guest, we would all look for her together [without commission]. ... When we all have guest, then we can go home.
>
> (Petra)

These cooperation arrangements brought people closer while making the community stronger and more resilient to the troubles they were facing. With each other's support and through mutual help, locals rose financially. Generally, they helped each other with inspections on Radunica and at the bus station; they would also help one another connect with new gatherers and hosts. This cooperative atmosphere even created new opportunities for those who were neither gatherers nor hosts but were financially struggling, like Jasmina who provides cleaning services for many local hosts:

> I worked at the market for almost 30 years ... And after 5 years of struggle, I lost the will for it ... then my friend told me there is one gentleman ... He needed a cleaning lady ... that's how I started. First with him and then it expanded.
> (Jasmina)

Through their joint work, many of the individuals in the area significantly improved their financial situation. The street started changing, dilapidated old houses were remodelled, gardens and the park were restored and, most importantly, people were self-employed and had a source of income. This outcome was a product of the individual and joint efforts of the locals who resisted various unfavourable circumstances.

Conclusion

Croatia's socio-economic situation and financial crisis nudged individuals on Radunica to engage first with paid-offline and then -online P2P-accommodation trading. This led to radical changes in their lives. They had to rely on the community and showed grit on an individual level. The individual financial crisis, which was caused by the post-war circumstances in Croatia motivated locals to engage in P2P-accommodation trading. In turn, this triggered the resilient behaviours in the community, resulting in cooperation networks as a survival tactic for the community. In this case study, motives depended on the low financial status of individuals, which was influenced by the external circumstances relating to the post-war crises and, therefore, vary from previous research on motives of renting private accommodation (Karlsson and Dolnicar, 2016; Farmaki and Stergiou, 2019).

We could argue that achieving community resilience on Radunica happened over three phases. The first phase was the acceptance of new circumstances, the second was the adoption of an affirmative outlook, and the third participation in the culture of cooperation or cooperation networks. The culture of cooperation was crucial for the resilience of the community although all three phases were imperative for individual resilience. Individuals had to be willing to participate in cooperation networks and ignore negative consequences. The resilient individuals were the ones who managed to: (i) build their own cooperation networks, professional tourism accommodation business and gain profit (i.e. became owners of hostels and networks of apartments); and (ii) succeed by their own criteria. Although most of the local community members of Radunica showed resilience in response to the financial crisis, the economic activity of offline P2P accommodation led to significant changes in the local culture and everyday life, transforming the nature of homes and reducing the numbers of local residents on Radunica (Tolić, 2019). This case study proves not only that online P2P-accommodation trading transforms the nature of homes, but rather that this can also happen when P2P-accommodation trading is offline.

The meaning of the term 'peer-to-peer accommodation' becomes broader, since the results show that on Radunica it did not mean strictly a 'host-to-tourist' way of doing business. For several years, this 'peer-to-peer' relationship was actually based on a triangular 'gatherer–host–tourist' cooperation, which has changed the culture of everyday life on the street for good. The findings offer a rare glimpse into the social relationships of a local community during the early stages of its coexistence with P2P accommodation. Given that Radunica today suffers from various negative impacts of overtourism, the research value is reflected in understanding the reaction of local people to this phenomenon. This could, in future, help prevent negative impacts of such activity on the local level thus enabling sociocultural sustainability as an important principle of sustainable tourism development.

References

Adger, W.N. (2000) Social and ecological resilience: are they related? *Progress in Human Geography* 24(3), 347–364.

Ahmad Fitri Amir, Ammar Abd Ghapar, Salamiah A. Jamal and Khairun Najiah Ahmad (2015) Sustainable tourism development: a study on community resilience for rural tourism in Malaysia. *Procedia - Social and Behavioural Sciences* 168, 116–122.

Barron, K., Kung, E. and Proserpio, D. (2020) The effect of home-sharing on house prices and rents: evidence from Airbnb. Posted on SSRN [Social Science Research Network], 5 March. DOI: 10.2139/ssrn.3006832

Bec, A., McLennan, C. and Moyle, B.D. (2016) Community resilience to long-term tourism decline and rejuvenation: a literature review and conceptual model. *Current Issues in Tourism* 19(5), 431–457. DOI: 10.1080/13683500.2015.1083538

Belk, R. (2014) Sharing versus pseudo-sharing in Web 2.0. *The Anthropologist* 18(1), 7–23.

Berkes, F. (2009) Evolution of co-management: role of knowledge generation, bridging organizations and social learning. *Journal of Environmental Management* 90(5), 1692–1702.

Breidbach, C.F. and Brodie, R.J. (2017) Engagement platforms in the sharing economy: conceptual foundations and research directions. *Journal of Service Theory and Practice* 27(4), 761–777.

Cheer, J.M., Milano, C. and Novelli, M. (2019) Tourism and community resilience in the Anthropocene: accentuating temporal overtourism. *Journal of Sustainable Tourism* 27(4), 554–572, DOI: 10.1080/09669582.2019.1578363

Coaffee, J. and Wood, D.M. (2006) Security is coming home: rethinking scale and constructing resilience in the global urban response to terrorist risk. *International Relations* 20(4), 503–517.

Croatian Bureau of Statistics (2020a) Towns in Statistics. Available at: https://www.dzs.hr/Hrv/publication/FirstRelease/results.asp?pString=Turizam&pSearchString=%Turizam% (accessed 17 December 2020).

Croatian Bureau of Statistics (2020b) Employees – an overview for the Republic of Croatia Employees – overview by counties. Available at: https://www.dzs.hr/Hrv/publication/FirstRelease/results.asp?pString=Zaposlenost%20iandpSearchString=%Zaposlenost%20i% (accessed 17 December 2020).

Dolnicar, S. (2019) A review of research into paid online peer-to-peer accommodation: launching the Annals of Tourism Research Curated Collection on peer-to-peer accommodation. *Annals of Tourism Research* 75, 248–264.

Farmaki, A. and Stergiou, D.P. (2019) Escaping loneliness through Airbnb host-guest interactions. *Tourism Management* 74, 331–333. DOI: 10.1016/j.tourman.2019.04.006

Farrell, B.H. and Twining-Ward, L. (2004) Reconceptualising tourism. *Annals of Tourism Research* 31(2), 274–295.

Folke, C., Hahn, T., Olsson, P. and Norberg, J. (2005) Adaptive governance of social–ecological systems. *Annual Review of Environment and Resources* 30(1), 441–473.

Gallopín, G.C. (2006) Linkages between vulnerability, resilience, and adaptive capacity. *Global Environmental Change* 16(3), 293–303.

Guest, G., Bunce, A. and Johnson, L. (2006) How many interviews are enough?: An experiment with data saturation and variability. *Field Methods* 18(1), 59–82.

Guest, G., Namey, E. and Chen, M. (2020) A simple method to assess and report thematic saturation in qualitative research. *PLoS ONE* 15(5): e0232076.

Gurran, N. and Phibbs, P. (2017) When tourists move in: how should urban planners respond to Airbnb? *Journal of the American Planning Association* 83(1), 80–92.

Habibi, M.R., Kim, A. and Laroche, M. (2016) From sharing to exchange: an extended framework of dual modes of collaborative non-ownership consumption. *Journal of the Association for Consumer Research* 1(2), 277–294.

Hardy, A. and Dolnicar, S. (2018) Types of network members. In: Dolnicar, S. (ed.) *Peer-to-peer Accommodation Networks: Pushing the Boundaries*. Goodfellow Publishers, Oxford, pp. 170–181.

Holladay, P.J. and Powell, R.B. (2013) Resident perceptions of social–ecological resilience and the sustainability of community-based tourism development in the Commonwealth of Dominica. *Journal of Sustainable Tourism* 21(8), 1188–1211.

Horn, K. and Merante, M. (2017) Is home sharing driving up rents? Evidence from Airbnb in Boston. *Journal of Housing Economics*, 38, 14–24.

Ikkala, T. and Lampinen, A. (2015) Monetizing network hospitality: hospitality and sociability in the context of Airbnb. In: *Proceedings of the 18th ACM Conference on Computer Supported Cooperative Work & Social Computing,* 15–18 March, Vancouver, Canada. ACM, New York, pp. 1033–1044.

Karlsson, L. and Dolnicar, S. (2016) Someone's been sleeping in my bed. *Annals of Tourism Research* 58, 159–162.

Kuhzady, S., Seyfi, S. and Béal, L. (2020) Peer-to-peer (P2P) accommodation in the sharing economy: a review. *Current Issues in Tourism* July (published online ahead of print). DOI: 10.1080/13683500.2020.1786505

Larsen, R.K., Calgaro, E. and Thomalla, F. (2011) Governing resilience building in Thailand's tourism-dependent coastal communities: conceptualizing stakeholder agency in social–ecological systems. *Global Environmental Change* 21(2) 481–491.

Lew, A.A. (2014) Scale, change and resilience in community tourism planning. *Tourism Geographies* 16(1), 14–22.

Lew, A.A. (2018) Planning for slow resilience in a tourism community context. In: Cheer, J. and Lew, A.A. (eds) *Tourism, Resilience and Sustainability: Adapting to Social, Political and Economic Change*. Routledge, Taylor & Francis Group, New York, pp. 34–59.

Lew, A.A, Ng, P.T., Ni, C-c. and Wu, T-c. (2016) Community sustainability and resilience: similarities, differences and indicators. *Tourism Geographies* 18(1), 18–27, DOI: 10.1080/14616688.2015.1122664

Linnenluecke, M.K., Griffiths, A. and Winn, M. (2012) Extreme weather events and the critical importance of anticipatory adaptation and organizational resilience in responding to impacts. *Business Strategy and the Environment* 21(1), 17–32.

Lutz, C. and Newlands, G. (2018) Consumer segmentation within the sharing economy: the case of Airbnb. *Journal of Business Research* 88, 187–196. DOI: 10.1016/j.jbusres.2018.03.019

Magis, K. (2010) Community resilience: an indicator of social sustainability. *Society & Natural Resources* 23(5), 401–416.

May, K. (2018) Airbnb officially opens up platform to hotel distribution. *Travel Weekly*, 7 February. Available at: https://www.travelweekly.com/Travel-News/Hotel-News/Airbnb-officially-opens-up-platform-to-hotel-distribution (accessed 7 July 2021).

McKercher, B. and Prideaux, B. (2020) *Tourism Theories, Concepts and Models*. Goodfellow Publishers, Oxford. DOI: 10.23912/9781911635352-4280

Norris, F.H., Stevens, S.P., Pfefferbaum, B., Wyche, K.F. and Pfefferbaum, R.L. (2008) Community resilience as a metaphor, theory, set of capacities, and strategy for disaster readiness. *American Journal of Community Psychology* 41(1–2), 127–150.

Oxford English Dictionary (2020a) Meanings of share in English. Available at: Share I Definition of Share by Oxford Dictionary on Lexico.com also meaning of Share (accessed 10 December 2020).

Oxford English Dictionary (2020b) Meaning of collaborate in English. Available at: https://www.lexico.com/definition/collaborate (accessed 10 December 2020).

Park, S.Y. and Agrusa, J. (2020) Short-term rental: disruptively innovative or disturbingly annoying for residents? *Asia Pacific Journal of Tourism Research* 25(8), 877–888. DOI: 10.1080/10941665.2020.1778049

Patel, S.S., Brooke Rogers, M., Amlôt, R. and Rubin, G.J. (2017) What do we mean by 'community resilience'? A systematic literature review of how it is defined in the literature. *PLoS Currents* 1(9).

Paton, D. and Johnson, D. (2001) Disasters and communities: vulnerability, resilience and preparedness. *Disaster Prevention and Management* 10(4), 270–277.

Piplović, S. (2006) Urbanistički i graditeljski razvitak predgrađa Lučac u Splitu. *Radovi Zavoda povijesnih znanosti HAZU u Zadru* 48, 453–510.

Reinhold, S. and Dolnicar, S. (2018a) The sharing economy. In: Dolnicar, S. (ed.) *Peer-to-peer Accommodation Networks: Pushing the Boundaries*. Goodfellow Publishers, Oxford, pp. 15–26.

Reinhold, S. and Dolnicar, S. (2018b) Airbnb's business model. In: Dolnicar, S. (ed.) *Peer-to-peer Accommodation Networks: Pushing the Boundaries*. Goodfellow Publishers, Oxford, pp. 27–38.

Reinhold, S. and Dolnicar, S. (2018c) How Airbnb creates value. In: Dolnicar, S. (ed.) *Peer-to-peer Accommodation Networks: Pushing the Boundaries*. Goodfellow Publishers, Oxford, pp. 39–53.

Reinhold, S. and Dolnicar, S. (2018d) How Airbnb captures and disseminates value. In: Dolnicar, S. (ed.) *Peer-to-peer Accommodation Networks: Pushing the Boundaries*. Goodfellow Publishers, Oxford, pp. 54–62.

Roelofsen, M. (2018) Performing 'home' in the sharing economies of tourism: the Airbnb experience in Sofia, Bulgaria. *Fennia-International Journal of Geography* 196(1), 24–42.

Schäfer, P. and Braun, N. (2016) Misuse through short-term rentals on the Berlin housing market. *International Journal of Housing Markets and Analysis* 9(2), 287–311.

Stabrowski, F. (2017) 'People as businesses': Airbnb and urban micro-entrepreneurialism in New York City. *Cambridge Journal of Regions, Economy and Society* 10(2), 327–347.

Stergiou, D.P. and Farmaki, A. (2020) Resident perceptions of the impacts of P2P accommodation: implications for neighbourhoods. *International Journal of Hospitality Management* 91: 102411. DOI: 10.1016/j.ijhm.2019.102411

Tolić, H. (2019) *Splitska Radunica kao turistička destinacija: studija društvenih interakcija*. Hrvatsko etnološko društvo, Ffpress, Zagreb.

Tsai, C.H., Wu, T-c., Wall, G. and Linliu, S-C. (2016) Perceptions of tourism impacts and community resilience to natural disasters. *Tourism Geographies* 18(2), 152–173. DOI: 10.1080/14616688.2016.1149875

World Bank Group (2018) *Tourism and the Sharing Economy: Policy & Potential of Sustainable Peer-to-Peer Accommodation*. The World Bank Group, Washington. Available at: http://documents1.worldbank.org/curated/en/161471537537641836/pdf/130054-REVISED-Tourism-and-the-Sharing-Economy-PDF.pdf (accessed 7 July 2021).

Zautra, A., Hall, J. and Murray, K. (2008) Community development and community resilience: an integrative approach. *Community Development* 39(3), 130–147.

5 Perceived Impacts of Short-term Rentals in the Local Community in the UK

Cristina Miguel,[1]* Christoph Lutz,[2] María del Mar Alonso-Almeida,[3] Brian Jones,[1] Filip Majetić[4] and Rodrigo Perez-Vega[5]
[1]University of Gothenburg, Sweden; [2]Norwegian Business School, Norway; [3]Autonomous University of Madrid, Spain; [4]Institute of Social Sciences Ivo Pilar, Croatia; [5]University of Kent, UK

Abstract
This study explores the perceived impacts of short-term rentals (STRs) in the UK through a quantitative analysis. A 57-item questionnaire covering sociocultural, economic, political, environmental and technological impacts of STRs was distributed online among guests, hosts and aware non-users. Apart from descriptive statistics, the data analysis included a principal component analysis to explore the dimensionality of the perceived impacts, and an analysis of variance (ANOVA) to assess differences in the community-related dimensions among the included groups. The results reveal that: (i) STRs are perceived relatively ambivalently but slightly more positively than negatively; (ii) environmental care, urban transformation, social capital and lifestyle, housing, infrastructure and events, and antisocial behaviour and crime are the key dimensions of the perceived impact; and (iii) stakeholder groups who were more involved in STRs perceived the impact of such rentals positively. We discuss the findings in the light of the literature on community resilience.

Keywords: community resilience, perceived impact, platforms, short-term rentals, urban regeneration

Highlights

- The chapter offers a quantitative assessment of the perceived impact of STRs in local UK communities.
- Overall, the respondents in this study perceived the impact of STRs slightly more positively than negatively.
- The perceived impact of STRs on urban regeneration can also improve community resilience by building on community identity and lifestyle.

Introduction

Companies such as Airbnb, HomeAway or Booking.com have been able to generate new short-term accommodation options for people who are visiting a specific area, without the need to create new assets or hire staff to provide the service. Such companies have helped deliver greater market efficiencies, by improving the amount and type of accommodation offered, and by reducing prices, both of which are factors that benefit the end user (Guttentag, 2015; Fang *et al.*, 2016).

*C.Miguel@leedsbeckett.ac.uk

(eds Anna Farmaki, Dimitri Ioannides and Stella Kladou)
DOI: 10.1079/9781789246605.0005

Additional economic benefits arise from complementary economies built around the influx of tourists using short-term rental (STR) platforms. These range from basic support services (e.g. cleaning, painting, repairing) to profit maximization services (e.g. dynamic pricing platforms) (Sigala, 2017). The emergence of these sub-economies, together with STRs themselves, contributes to employment opportunities, increased tax revenues and supports tourism in general (Forgacs and Dolnicar, 2017; DiNatale *et al.*, 2018).

By contrast, by removing property from the long-term rental market, STRs can lead to increased rents as supply and demand diverges (Guttentag, 2015; Barron *et al.*, 2017; Slee, 2017). Moreover, neighbourhoods are transformed and gentrified with the arrival of tourists, so that long-term tenants might be displaced (Wachsmuth and Weisler, 2018). STRs might create new gentrification or exacerbate ongoing gentrification that has preceded tourism development. Other negative impacts relate to the disruption that visitors using STRs generate, namely: (i) noise which leads to sleep deprivation (Gant, 2016); (ii) frequency of tenant turnover (Smorto, 2016); and (iii) the transformation of the character of the neighbourhood (Ranchordas *et al.*, 2016). Further, reports of tourists trying to open residents' doors because they do not know where their apartment is (Gant, 2016) raise concerns about public safety and interpersonal privacy (Ranchordas *et al.*, 2016; Lutz *et al.*, 2018). Cócola-Gant and Gago (2019) frame their case study of Lisbon from a critical political economic analysis that locates the growth of STRs within the circuit of capital. However, limited research has looked into the perceived impact of STRs (Jordan and Moore, 2018; Mody *et al.*, 2019; Stergiou and Farmaki, 2019) and the role of STRs and their perceived impact in relation to community resilience is unclear. Do STRs and their impact enhance community resilience or disrupt communities by making them more divided and less resilient? To address these gaps and add more quantitative evidence on the topic, we conducted a survey on perceived impacts of STRs among UK-based users (hosts and guests) and non-users who are aware of the STR platforms.

The chapter is organized into five parts. After this introduction, we provide a short overview of the literature on the perceived impacts of STR platforms on local communities and introduce the concept of community resilience as a conceptual lens through which we approached the research described in this chapter. In the third part, we present the methodology and in the fourth part, an overview of the results. Finally, in the discussion and conclusion section the key findings are summarized and discussed in the light of the literature on community resilience.

Theory

In comparison with the actual impacts of STRs (Guttentag, 2015; Fang *et al.*, 2016; Barron *et al.*, 2017; Forgacs and Dolnicar, 2017; Sigala, 2017; Slee, 2017; DiNatale *et al.*, 2018; Wachsmuth and Weisler, 2018), perceived impacts have not yet been investigated in depth (Mody *et al.*, 2020). Within STRs, peer-to-peer (P2P) accommodation is more than merely an economic exchange since it also includes a social dimension (Richardson, 2015). STRs, especially in P2P form, come with moral and ethical questions (Farmaki *et al.*, 2019) that shape and inform community perceptions of impact. Measures of perceived impact on communities should ideally include an ethical dimension that addresses issues such as justice and rights (Ahsan, 2020). Consideration also needs to be given to the sustainability of the P2P-accommodation model and its impact on the natural environment in which communities are located (Seravalli, 2014). Stergiou and Farmaki (2019) conducted qualitative research in Athens to investigate residents' perception of P2P-accommodation impacts on their neighbourhood. The authors found that the positive impacts, such as economic benefits to hosts, increased socializing opportunities for hosts offering home-sharing (room rental), and revival of the 'overly quiet' neighbourhood were (heavily) outweighed by the negative impacts. Examples of these included: (i) lack of housing units; (ii) overall increased level of noise; (iii) increased amount of litter on the streets left by travellers/guests; (iv) privacy and security issues; (v) tax avoidance; (vi) 'communal area damages'; and (vii) decreased sense of community.

The study of perceived impacts, in addition to actual impacts, is important because perception shapes reality. More concretely, citizens might not be fully aware of the actual impacts of STRs but still have opinions based on folk theories, their own experiences with such services and their conversations with other people about such phenomena. Nieuwland and van Melik (2020) found that both perceived and actual impacts are important factors that determine citizens' attitudes and regulatory approaches towards STRs. From a practical point of view, these perceptions can directly affect different stakeholders, including: (i) STR platforms (which might attract fewer users if the perceptions are largely negative); (ii) policy makers (who might be more restrictive and quicker to regulate if they see platforms acting unfairly; see Newlands and Lutz, 2020); and (iii) communities. Local communities, for example, could face more tensions if perceived impacts are negative or very mixed. In that regard, a community's resilience is of importance.

Community resilience is a contested and much discussed topic in academic literature (e.g. Erdmenger, 2019), policy, and practice (Elstow, 2013). It has been explored and applied through a number of subject disciplinary lenses (Koliou *et al.*, 2020). Community resilience seeks to create capability and capacity by enabling, enriching and empowering by broadening relations, accepting diverging positions and facilitating inclusion. It is an academic construct in search of real-world application and, as such, it takes varied forms according to the interpretation given, the circumstances in which it operates, and the stakeholders to which it is applied (Elstow, 2013). Community resilience seeks to develop opportunities that strengthen ties within and between communities that allow them to overcome challenges and be stronger together. Some early evidence suggests that in the presence of crisis and various disasters (e.g. weather related, earthquakes), P2P-accommodation networks emerge to support displaced tourists as the traditional hospitality infrastructure is damaged (Hajibaba *et al.*, 2017; Hajibaba and Dolnicar, 2018a). Furthermore, Volgger *et al.* (2018) emphasize that stress factors are cumulative and that tourism destinations are being challenged and, therefore, need the capabilities to adapt. Destination resilience is reliant on issues of governance, response to that and the ability to adapt (Volgger *et al.*, 2018). STRs make both a negative and a positive contribution to destination resilience (Volgger *et al.*, 2018). Erdmenger (2019, p. 437) refers to 'the tolerance and resilience of stressed communities' and uses the concept of community resilience to explore destination communities in the context of increasing numbers of tourists.

The ability to organize, influence, shape and change events, agendas, interventions, policy and practice are key to resilient communities and help them adapt to as well as withstand shocks that might otherwise break them (Murphy, 2007; Pike *et al.*, 2010). Community resilience comprises different elements according to the situation to which it is applied or discussed. Different stakeholders claim aspects of it as their own and seek to advance their cause under the remit provided by the umbrella term community resilience. Transformational change can be delivered as part of community resilience, and this typically comes about through social learning (Wilson, 2012). The concept of resilience resides both at individual and at collective societal levels and is very much about the ability to adapt to extreme environmental constraints or strained circumstances, for example the disruptive nature of STRs on local communities. The ways in which communities learn, acquire and develop resilience and deal with unexpected, unforeseen challenges is a common theme (Buikstra *et al.*, 2010; Kulig *et al.*, 2013). Resilience thinking is key to building stronger, more adaptable, enabling communities that can meet future challenges robustly and with confidence (Berkes and Ross, 2013). We use the lens of community resilience to discuss the perceived impacts of STRs in more depth below.

Methodology

Our study sample included UK-based participants throughout the country who were reached via an online questionnaire administered through Amazon Mechanical Turk (MTurk) in July 2019. MTurk is a marketplace for easy access to a diverse, on-demand workforce in exchange for payment. It has been identified as a useful resource for data collection in social sciences due to its affordability, simplicity and scalable accessibility to human intelligence. For this reason, MTurk is widely used in consumer research (Goodman and Paolacci, 2017; Aguinis *et al.*, 2020). Some reliability and validity concerns around the use of MTurk to collect data for research have been raised, although tests have shown that data provided by MTurk is as reliable as data collected through other means (Casler *et al.*, 2013; Thomas and Clifford, 2017). Since this study aimed to measure perceived impacts of STR platforms in the entire population, no restrictions were introduced, meaning that the data collection aimed at both users of these platforms (guests and/or hosts) and non-users who are aware of them. In total 138 responses where received, of which 120 responses were valid and used for further analysis.

To ensure comprehensiveness of the impact assessment, the 57-item questionnaire included the sociocultural, economic, political, environmental and technological impacts that STRs could cause. Due to the lack of previous attempts to (quantitatively) explore perceived impacts of STRs and tourism in general, the dimensions of impacts were not represented by predefined components/factors but composed of individual items adapted predominantly from Andereck *et al.* (2005), Hammar *et al.* (2009), Liang *et al.* (2018) and Mody *et al.* (2019). As an illustration, in Mody *et al.*'s study (2019) on positive and negative impacts of Airbnb the item assessing the extent to which Airbnb 'fosters community pride' indicates a sociocultural impact, the extent to which Airbnb 'provides more business for local people and small businesses' indicates an economic impact, while perceived 'increase in the number of Airbnb visitors that results in noise and pollution/litter' might indicate an environmental impact. Furthermore, Andereck *et al.* (2005) investigated residents' perceptions of tourism's multidimensional impact on communities. Their dimensions of community environment (e.g. peace and quiet), community problems (e.g. crowding and congestion) and community life (e.g. an understanding of different cultures) represent sociocultural and environmental impacts while the dimensions of community services (e.g. fair prices for goods and services) and community economy (e.g. the strength and diversity of the local economy) might be more related to economic impacts. In addition, we developed eight items ourselves (offering of STRs, demand of STRs, attractions, bohemian environment, melting pot environment, number of houses sold per square kilometre, working conditions in my city, hidden/black economy practices). These items were to cover community-related aspects which had not been addressed in the questionnaires we predominantly relied on, but were also found relevant in the literature review (e.g. DiNatale *et al.*, 2018; Nieuwland and van Melik, 2020). All 57 items, as they were asked in the survey, are listed in the Appendix at the end of this chapter. The survey participants had six response options for all impact items: 1 – no impact; 2 – very negative impact; 3 – somewhat negative impact; 4 – ambivalent impact; 5 – somewhat positive impact; and 6 – very positive impact. The decision to include a no impact category, which is treated distinctly from the valence-based response options (2–6), was made by the research team based on deliberations that a perceived lack of impact is different from a perceived ambivalent impact. The latter category means that someone perceives both negative and positive aspects and that the positive and negative impacts are roughly equally pronounced. By contrast, a perceived lack of impact (no impact) means that someone thinks that STRs and the phenomenon presented in the item are not causally associated. In the analyses reported below, we excluded 'no impact' responses by treating them as missing values when we were interested in the directionality and comparison of individual items. While we did not directly measure community resilience in the survey, several of our items refer to community aspects and can be interpreted within a community resilience perspective (e.g. 'a feeling of belonging in my community', 'regeneration of neighbourhoods in decline').

In terms of data analysis, in addition to descriptive statistics, principal component analysis (PCA) was performed to assess the dimensionality of the questionnaire. We considered only components with eigenvalues of one or higher and the axes were rotated using Varimax with Kaiser normalization. Only factor loadings over 0.5 were used and we eliminated items with high cross-loadings. Finally, an analysis of variance (ANOVA) was performed to assess differences in the community-related dimensions of perceived impacts among the included groups. All analyses were conducted in IBM SPSS Statistics (v.25).

Findings

The results are shown in three subsections. First, we present and discuss descriptive statistics and general tendencies in the data. Second, we investigate the dimensionality of the perceived impacts. Third and finally, we focus on two of the five dimensions that described community aspects (social capital and lifestyle, antisocial behaviour and crime) to connect our results more directly to the theme of community resilience by analysing perception differences between guests, hosts, guests and hosts, and non-users.

Descriptive statistics and general tendencies

Across all the 57 perceived impact items (see Appendix), between 8% and 33% of respondents chose the 'No impact' option (code 1, see previous section). 'The preservation of my way of life' had the highest proportion of respondents choosing the 'No impact' option (33%), while 'hotel demand' had the lowest proportion (8%). Respondents felt that STRs were least impactful on their own lifestyle (way of life) but most impactful for hotels. Thus, respondents see a clear impact of STRs on various aspects of their neighbourhood, particularly so when it comes to the perceived impact on competitors (hotels). In a next step, we looked into whether these perceptions are positive or negative on aggregate. For this analysis, we excluded respondents who reported no perceived impact from the analysis, as they would have affected the polarity of impact analysis, no matter the coding.

Across all items, the arithmetic mean was 3.24 (on a 1–5 scale, where: 1 – very negative impact; 2 – somewhat negative impact; 3 – ambivalent impact; 4 – somewhat positive impact; 5 – very positive impact). Thus, respondents see the impact of STRs relatively ambivalently but slightly more positively than negatively. This contrasts with previous studies, which often stressed the negative impacts (Jordan and Moore, 2018; Stergiou and Farmaki, 2019). However, the differences in findings could be due to the different methodology (quantitative vs qualitative), the different sampling approach and the study location.

The arithmetic means ranged from 2.48 (for the 'crowding and congestion' item) and 2.63 (for two items, namely 'traffic issues' and 'drug and alcohol use') on the low end of the spectrum to 3.75 (for the 'employment' item) and 3.72 (for the 'ethnic diversity' item) on the high end. We also noted a relatively negative impact on 'litter issues' (2.69), 'conflicts over zoning/land use' (2.71), and 'rents' (2.84). By contrast, comparatively positive impacts occurred for 'festivals, fairs and museums' (3.71) and 'income' (3.65). Thus, STRs are perceived as exacerbating problematic aspects, while enhancing positive and beneficial aspects. The negative aspects relate to disruptions of neighbourhoods and describe social and environmental concerns, whereas the positive aspects mostly relate to economic benefits that occur from STRs. These findings are in line with evidence from qualitative studies (Jordan and Moore, 2018; Stergiou and Farmaki, 2019).

Overall, while our findings align with previous research in terms of aspects identified as positive and negative, differences appear in the overall perception. However, our strategy to assess overall perception as an aggregation of individual perceptions does not do justice to the fact that individuals might weigh the different aspects differently. For example, while someone might see very positive employment effects of STRs based on their knowledge of the topic through media coverage, the immediate disruptive effects they experience in their community might largely outweigh the positive perceptions, leading to an aggregate negative impact assessment. Unfortunately, we did not include a question to assess the overall perceived impact of STRs, as the focus was on differentiated impacts and their dimensionality. However, the aggregate score listed above across all items (3.24) can be interpreted as a summed-up perceived impact score that captures different impact dimensions and indicates an ambivalent to slightly positive tendency. This aligns with Mody *et al.* (2019), who investigated positive and negative perceived impacts separately, rather than as two ends of the spectrum. Their 14 positive perceived impacts items had an arithmetic mean of 3.62 on a 1–5 scale and their 15 negative perceived impacts items had an arithmetic mean of 2.78. The weighted average of their perceived impacts items is 3.19 and thus quite close to our value, also showing a net positive perceived impact. Mody *et al.* (2019) concluded that 'residents perceive higher positive than negative impacts of Airbnb, a finding that is the opposite of what is often portrayed in the media' (p. 200).

Dimensionality of impacts

The PCA yielded five components encompassing 29 items in total (see Table 5.1). Varimax rotation converged in six iterations. Based on the item content, we labelled the five components as follows: (i) environmental care (EC); (ii) urban transformation (UT); (iii) social capital and lifestyle (SCL); (iv) housing, infrastructure and events (HIE); and (v) antisocial behaviour and crime (ABC). Of the total variance, 69.36% is explained with the five factors and the Kaiser-Meyer-Olkin (KMO) value being 0.912, indicating 'marvellous' fit (Kaiser, 1974, p. 35). All components had satisfactory internal consistency, as indicated by the Cronbach's α values, which exceeded the commonly accepted threshold of 0.70 (Nunnally, 1978): 0.89 for EC, 0.85 for UT, 0.86 for SCL, 0.71 for HIE and 0.82 for ABC.

The first component, EC, has nine items. All items relate to environmental or sociocultural sustainability. The three items with the highest factor loading values are about preservation, either of natural or cultural heritage. The six remaining items describe basic residential conditions, particularly the presence or absence of polluting factors such as traffic, litter and noise.

The second component, UT, has six items. The first three items revolve around gentrification and the sociocultural impact of STRs on neighbourhoods. Related to that, 'hidden/black economy practices' and 'working conditions in my city' capture potential economic effects that are associated with urban renewal and transformation of neighbourhoods. Finally, the number of houses sold per square kilometre is an economic indicator that captures the dynamics of the housing market, which is also associated with urban transformation and gentrification.

The third component, SCL, has seven items. They describe sociocultural and community-related aspects. The first three items and the sixth item revolve around how STRs affect someone's way of life within a given community, either directly or indirectly through tourism. The remaining three items are about political and social implications resulting from the presence of STRs. They seem to be more on an aggregate level, capturing aspects such as urban development, while the items about someone's way of life are more individualistic.

The fourth component, HIE, is about the availability of short-term housing more generally, thus aligning with discussions about potential crowding out of local populations through STR platforms. However, the component also includes items on adjacent topics such as attractions and public transport. This might imply that, in the perception of the respondents, STR availability in general is strongly tied to other infrastructural issues.

The fifth and final component, ABC, only has two items. Both items describe mostly negatively associated phenomena in a given neighbourhood. The fact that this component is distinct from others shows that it might pose a particularly salient and differentiated impact dimension among the respondents.

The PCA shows that the perceived impacts can be grouped into relatively separate dimensions that span sustainability-related and environmental aspects (EC), the specific living conditions of urban environments (UT), the interplay of individuals and their community (SCL), infrastructural concerns centred around economic aspects and amenities (HIE), and some dark sides (ABC). Sustainability is conceptually defined through its three sub-components, which refer to social, environmental and economic/financial sustainability (Elkington, 1998). Yet, our data analysis does not reflect such a clear distinction among the three sub-components. This shows how the perceptions of STRs reflect a lived reality where cultural, social, economic and environmental perceptions are heavily intertwined. For example, the mix of items in SCL indicates how population-related (urban sprawl) and political questions (resident participation on local government) are strongly entangled with seemingly more individualistic aspects (my personal life quality; the preservation of my way of life). It also shows the importance of seeing STRs from an institutional and power-related perspective, where platforms such as Airbnb are increasingly involved in urban governance, shaping how cities are transformed (Van Doorn, 2020).

Community-related aspects and group differences

In the following paragraphs, we focus on the community-related dimensions, SCL and ABC. We compared them across the different user groups in

Table 5.1. Perceived impacts of short-term rentals (STRs).[a]

	Dimensions describing community aspects[b]				
Impact item	EC	UT	SCL	HIE	ABC
The preservation of wildlife habitats	0.743				
The preservation of cultural/historical sites	0.734				
The preservation of natural areas	0.733				
Clean air and water	0.689				
Traffic issues	0.660				
Peace and quiet	0.659				
Litter issues	0.638				
Crowding and congestion	0.623				
Quality recreation opportunities	0.615				
Melting pot environment		0.754			
Bohemian environment		0.730			
Regeneration of neighbourhoods in decline		0.716			
Hidden/black economy practices		0.704			
Working conditions in my city		0.653			
Number of houses sold per square kilometre		0.562			
A feeling of belonging in my community			0.708		
My personal life quality			0.677		
The preservation of my way of life			0.662		
Urban sprawl and population growth			0.655		
Resident participation in local government			0.645		
Having tourists who respect my way of life			0.615		
Having live sports to watch in my community			0.582		
Demand of STRs in general				0.777	
Offering of STRs in general				0.753	
Festivals, fairs and museums				0.741	
Good public transportation				0.595	
Attractions				0.561	
Crime and vandalism					0.845
Drug and alcohol abuse					0.818

[a]Values shown are factor loadings obtained from principal component analysis (PCA) (see Methodology). Only factor loadings over 0.5 were used. A high factor loading value indicates that the impact of that item is considered to be greater than an item with a lower factor loading value.
[b]EC = Environmental care; UT = Urban transformation; SCL = Social capital and lifestyle; HIE = Housing, infrastructure and events; ABC = Antisocial behavior and crime. N = 120.

our sample: (i) guests; (ii) hosts; (iii) guests and hosts; and (iv) non-users. For SCL, according to the performed ANOVA, we found statistically significant differences at the 5% level ($F = 2.908$, p value = 0.038). Hosts had the highest component score (0.78 on the z-standardized SCL score), followed by guests and hosts (i.e. respondents who are using STR platforms as both hosts and guests). This group had a component score of 0.45. While the non-user score was close to the arithmetic mean of 0 (0.11), guests had the lowest scores (-0.18). The findings indicate that the 'more involved' groups (hosts; guest and hosts) have the most positive perception of STR in terms of social capital and lifestyle. It could be that these groups experience the community-related benefits of STRs most strongly, while guests and non-users have a more transactional (in the case of guests) or outsider (in the case of non-users) perspective, where the community-related benefits are less directly felt.

When it comes to ABC, we did not find statistically significant perception differences between the groups in the ANOVA ($F = 1.230$, $p = 0.302$). Comparing the component scores, hosts again scored highest (0.88), followed by guest and hosts (0.23), guests (-0.10) and non-users (-0.56). This indicates that non-users see STR as most negative in terms of ABC. Again, hosts and guest and hosts might overlook these aspects more strongly. As a caveat, the host category only includes two respondents (most hosts are also guests, with the guest and hosts category having 24 respondents). However, grouping the guest and hosts category within hosts (with the rationale the more involved category of hosts should be the main category) does not change the results. The STR differences remain significant at the 5% level (F = 4.285, p = 0.016), while the ABC differences remain insignificant (F = 1.461, p = 0.235).

Looking at individual perception differences and using the three-category grouping (guest, host, non-user; the 24 guests and hosts are categorized as hosts and merged with the two hosts), we also find pronounced differences for individual items. Hosts think that STRs create a feeling of belonging more strongly than guests and non-users (4.55 for hosts vs 4.06 for guests and 3.82 for non-users); also that STRs improve their life quality (4.95 vs 4.36 and 3.82, respectively), allow them to preserve their way of life (4.95 vs 4.15 and 4.00, respectively), and lead to more resident participation in local government (4.77 vs 4.09 and 3.73, respectively). For all the seven SCL items, hosts score most positively, while non-users tend to score more negatively. Hosts also have the most positive perception how STRs reduce vandalism and crime (4.18 for hosts vs 3.91 for guests vs 3.38 for non-users) as well as drug and alcohol abuse (4.14 vs 3.68 vs 3.19, respectively). These findings seem somewhat expected because hosts, influenced by their own commercial interests, tend to promote STRs as beneficial to the community's quality of life. Conversely, non-users' perception could be influenced by negative news about STRs more than in their own experience. Unfortunately, there exists little research into the perceived impacts of STRs more generally, let alone from a comparative perspective that contrasts different user groups and non-users. However, Mody *et al.* (2020) showed how psychological and social empowerment are key drivers of the perceived impacts of Airbnb among non-hosts in the USA, with those feeling more politically empowered perceiving Airbnb's impacts more negatively. Residents who feel politically empowered and feel particularly negatively affected by STRs were quite active in organizing against overtourism, including STRs (Ramos and Mundet, 2020). Beyond perceived impacts, however, some research has compared consumers (i.e. guests), providers (i.e. hosts) and non-users (Andreotti *et al.*, 2017; Angelovska *et al.*, 2020) in terms of their motives to participate in the sharing economy. The findings indicate that providers have higher values for non-economic motives than consumers, whereas consumers score highest on economic motives. Given that non-economic motives reflect a more community-oriented approach, these findings align with our results that hosts perceive the impacts of STRs most positively.

Discussion and Conclusion

Most of the investigated perceived impacts of STRs tend to skew positively for the community, rather than negatively. This is especially true for environmental aspects and historical destinations. STR platforms could help avoid some of the problems of mass tourism, including sun and beach tourism and cruise tourism, allowing a better management of touristic flows and helping to break seasonality. Therefore, although the real impacts of STRs are not conclusive, these findings provide insights to understand the phenomenon of STRs and their impacts on community resilience that will be necessary to be deepened in the future. First, our data showed different ways in which respondents believed STRs can contribute to urban regeneration. STR platforms can help promote new areas in the city, even neighbourhoods in decline (Aguilera *et al.*, 2019; Alonso-Almeida *et al.*, 2019). Thus, STRs, especially if they are P2P-oriented, could reinforce community resilience in the sense that they help to transform deteriorating areas and turn them into new spaces for life. For example, they could include systems of valuation of the neighbourhood to assess the environmental quality and how it evolves over time. Second, STRs can provide a source of wealth for the city. They create new jobs, increase local taxes, promote innovation and reinforce local small companies. By contrast, positive perceptions of SCL (social capital and lifestyle) aspects also contribute to community resilience by preserving different lifestyles and boosting the community identity. This last point is very relevant given that community ties make the whole destination stronger and facilitate destination governance.

Our study findings showed that the respondents perceived STRs slightly more positively than negatively on aggregate. Nevertheless, as seen earlier, the dimensions capture both positive and negative perceptions, with the exception of ABC (antisocial behaviour and crime), which is clearly negative. Therefore, it is difficult to separate positive and negative perceptions very clearly. Our findings showed differences with previous qualitative research, which emphasized the negative perception of STR platforms (e.g. Stergiou and Farmaki, 2019) but these were mostly aligned with quantitative research that showed ambivalence to positive perceived impacts on aggregate (Mody *et al.*, 2019). Moreover, in previous research positive and negative perceptions of tourism were clearly separated by domains (see for example Andereck *et al.*, 2005). Nevertheless, those findings pointed out that deeper changes are emerging as a result of STR platforms' penetration in cities and tourist destinations. Thus, new dominions (or dimensions) emerge with both positive and negative impacts.

Also, building on previous work on STRs and their sustainability (see Midgett *et al.*, 2018; Ceptureanu *et al.*, 2020), EC (environmental care) was found to be perceived as one of the more positive impacts derived from STR platforms in our study. Despite some early evidence of the actual impact of STRs on the environment (Cheng *et al.*, 2020), our findings suggest that at a local level, the perceived impact of STR platforms as collaborative consumption are aligned with the view that sharing resources is a main driver to improve environmental issues. Obviously, the more intensive use of certain accommodation places could increase the total amount of litter produced and water consumption and/or reduce the enjoyment of natural resources. However, by contrast, proper regulation of touristic flows and political measures associated with the bigger numbers of visitors can contribute to improve both natural or cultural heritage and basic residential conditions and make clearer the role of STR platforms in sustainability. In this sense, Duignan (2019) proposed to develop strategies based on the definition of the carrying capacity of the city and of critical areas and attractions, such as monitoring all operators and touristic flows and reviewing regulation on access to certain areas of the city for tourist-related activities.

The ABC dimension was found to be perceived as the worst impact of STRs. Thus, governments should pay attention to a possible increase in crime, delinquency or other antisocial behaviours in most popular STR neighbourhoods in order to make arrangements that produce feelings of safety for both residents and visitors. A concrete recommendation in this context could be to make the creation of an STR register mandatory by law. This is something that already happens in cities such as Barcelona, Paris, San Francisco and Tokyo (Hajibaba and Dolnicar, 2018b). This measure gives security to visitors and contributes to restricting illegal activities.

To sum up, the increase in numbers of tourists can contribute to positive changes in cities (Alonso-Almeida *et al.*, 2019), but policy makers should regulate STRs adequately to prevent negative impacts as much as possible and to enhance positive ones. This could lead to more favourable perceptions and higher community resilience in the long run. Thus, a future line of research should analyse in-depth impacts of STRs on communities and their contribution to community prosperity and crisis situation resilience.

References

Aguilera, T., Artioli, F. and Colomb, C. (2019) Explaining the diversity of policy responses to platform-mediated short-term rentals in European cities: a comparison of Barcelona, Paris and Milan. *Environment and Planning A: Economy and Space*. DOI: 10.1177/0308518X19862286

Aguinis, H., Villamor, I. and Ramani, R.S. (2020) MTurk research: review and recommendations. *Journal of Management* 19 November, 1–15. DOI: 10.1177/0149206320969787

Ahsan, M. (2020) Entrepreneurship and ethics in the sharing economy: a critical perspective. *Journal of Business Ethics* 161, 19–33.

Alonso-Almeida, M.M., Borrajo-Millán, F. and Yi, L. (2019) Are social media data pushing overtourism? The case of Barcelona and Chinese tourists. *Sustainability* 11(12), 3356.

Andereck, K.L., Valentine, K.M., Knopf, R.C. and Vogt, C. (2005) Residents' perceptions of community tourism impacts. *Annals of Tourism Research* 32(4), 1056–1076.

Andreotti, A., Anselmi, G., Eichhorn, T., Hoffmann, C.P., Jürss, S. and Micheli, M. (2017) European perspectives on participation in the sharing economy. SSRN [Social Science Research Network], 3 October. Available at: https://papers.ssrn.com/sol3/papers.cfm?abstract_id=3046550 (accessed 12 December 2020).

Angelovska, J., Čeh Časni, A. and Lutz, C. (2020) Turning consumers into providers in the sharing economy: exploring the impacts of demographics and motives. *Ekonomska Misao i Praksa* 29(1), 79–100.

Barron, K., Kung, E. and Proserpio, D. (2017) The effect of home-sharing on house prices and rents: evidence from Airbnb. SSRN [Social Science Research Network], 25 July. Available at: https://ssrn.com/abstract=3006832 (accessed 12 December 2020).

Berkes, F. and Ross, H. (2013) Community resilience: toward an integrated approach. *Society and Natural Resources* 26(1), 5–20.

Buikstra, E., Ross, H., King, C.A., Baker, P.G., Hegney, D., *et al.* (2010) The components of resilience: perceptions of an Australian rural community. *Journal of Community Psychology* 38(8), 975–991.

Casler, K., Bickel, L. and Hackett, E. (2013) Separate but equal? A comparison of participants and data gathered via Amazon's MTurk, social media, and face-to-face behavioral testing. *Computers in Human Behavior* 29, 2156–2160.

Ceptureanu, E.G., Ceptureanu, S.I., Herteliu, C. and Cerqueti, R. (2020) Sustainable consumption behaviours in P2P accommodation platforms: an exploratory study. *Soft Computing* 24, 13863–13870. DOI: 10.1007/s00500-020-04681-6

Cheng, M., Chen, G., Wiedmann, T., Hadjikakou, M., Xu, L. and Wang, Y. (2020) The sharing economy and sustainability – assessing Airbnb's direct, indirect and induced carbon footprint in Sydney. *Journal of Sustainable Tourism* 28(8), 1083–1099.

Cócola-Gant, A. and Gago, A. (2019) Airbnb, buy-to-let investment and tourism-driven displacement: a case study in Lisbon. *Environment and Planning A: Economy and Space* 19 August, 1–18. DOI: 10.1177/0308518X19869012

DiNatale, S., Lewis, R. and Parker, R. (2018) Short-term rentals in small cities in Oregon: impacts and regulations. *Land Use Policy* 79, 407–423.

Duignan, M. (2019) 'Overtourism'? Understanding and managing urban tourism growth beyond perceptions: Cambridge case study: strategies and tactics to tackle overtourism. In: World Tourism Organization; Centre of Expertise Leisure, Tourism & Hospitality; NHTV Breda University of Applied Sciences; and NHL Stenden University of Applied Sciences (eds) *'Overtourism'? Understanding and Managing Urban Tourism Growth beyond Perceptions: Case Studies*. United Nations World Tourism Organisation (UNWTO), Madrid, pp. 34–39.

Elkington, J. (1998) *Cannibals with Forks: the Triple Bottom Line of 21st Century Business*. New Society Publishers, Gabriola, British Colombia.

Elstow, L. (2013) *Beyond Z-Cards and Grab Bags: Community Resilience in Urban Communities. Final Report for Winston Churchill Memorial Trust*. Winston Churchill Memorial Trust, London.

Erdmenger, E. (2019) Community resilience in urban tourist destinations: how beer garden romance and a hygge localhood boost social capital. *Zeitschrift für Tourismuswissenschaft* 11(3), 437–450.

Fang, B., Ye, Q. and Law, R. (2016) Effect of sharing economy on tourism industry employment. *Annals of Tourism Research* 57, 264–267.

Farmaki, A., Stergiou, D.P. and Kaniadakis, A. (2019) Self-perceptions of Airbnb hosts' responsibility: a moral identity perspective. *Journal of Sustainable Tourism*, 1–21. DOI: 10.1080/09669582.2019.1707216

Forgacs, G. and Dolnicar, S. (2017) The impact on employment. In: Dolnicar, S. (ed.) *Peer-to-peer Accommodation Networks: Pushing the Boundaries*. Goodfellow Publishers, Oxford, pp. 160–169.

Gant, A.C. (2016) Holiday rentals: the new gentrification battlefront. *Sociological Research Online* 21(3), 1–9.

Goodman, J.K. and Paolacci, G. (2017) Crowdsourcing consumer research. *Journal of Consumer Research* 44(1), 196–210.

Guttentag, D. (2015) Airbnb: disruptive innovation and the rise of an informal tourism accommodation sector. *Current Issues in Tourism* 18(12), 1192–1217.

Hajibaba, H. and Dolnicar, S. (2018a) Helping when disaster hits. In: Dolnicar, S. (ed.) *Peer-to-peer Accommodation Networks: Pushing the Boundaries*. Goodfellow Publishers, Oxford, pp. 237–245.

Hajibaba, H. and Dolnicar, S. (2018b) Regulatory reactions around the world. In: Dolnicar, S. (ed.) *Peer-to-peer Accommodation Networks: Pushing the Boundaries*. Goodfellow Publishers, Oxford, pp. 120–136.

Hajibaba, H., Karlsson, L. and Dolnicar, S. (2017) Residents open their homes to tourists when disaster strikes. *Journal of Travel Research* 56(8), 1065–1078.

Hammar, H., Jagers, S.C. and Nordblom, K. (2009) Perceived tax evasion and the importance of trust. *The Journal of Socio-Economics* 38(2), 238–245. DOI: 10.1016/j.socec.2008.07.003
Jordan, E.J. and Moore, J. (2018) An in-depth exploration of residents' perceived impacts of transient vacation rentals. *Journal of Travel & Tourism Marketing*, 35(1), 90–101.
Kaiser, H.F. (1974) An index of factorial simplicity. *Psychometrika* 39(1), 31–36.
Koliou, M., van de Lindt, J.W., McAllister, T.P., Ellingwood, B.R., Dillard, M. and Cutler, H. (2020) State of the research in community resilience: progress and challenges. *Sustainable and Resilient Infrastructure* 5(3), 131–151.
Kulig, J.C., Edge, D.S., Townshend, I., Lightfoot, N. and Reimer, W. (2013) Community resiliency: emerging theoretical insights. *Journal of Community Psychology* 41(6), 758–775.
Liang, L.J., Choi, H.C. and Joppe, M. (2018) Exploring the relationship between satisfaction, trust and switching intention, repurchase intention in the context of Airbnb. *International Journal of Hospitality Management* 69, 41–48.
Lutz, C., Hoffmann, C.P., Bucher, E. and Fieseler, C. (2018) The role of privacy concerns in the sharing economy. *Information, Communication & Society* 21(10), 1472–1492.
Midgett, C., Bendickson, J.S., Muldoon, J. and Solomon, S.J. (2018) The sharing economy and sustainability: a case for Airbnb. *Small Business Institute Journal* 13(2), 51–71.
Mody, M., Suess, C. and Dogru, T. (2019) Not in my backyard? Is the anti-Airbnb discourse truly warranted? *Annals of Tourism Research* 74, 198–203.
Mody, M., Woosnam, K.M., Suess, C. and Dogru, T. (2020) Hapless victims or empowered citizens? Understanding residents' attitudes towards Airbnb using Weber's Theory of Rationality and Foucauldian concepts. *Journal of Sustainable Tourism*, 1–23. DOI: 10.1080/09669582.2020.1834567
Murphy, B.L. (2007) Locating social capital in resilient community-level emergency management. *Natural Hazards* 41(2), 297–315.
Newlands, G. and Lutz, C. (2020) Fairness, legitimacy and the regulation of home-sharing platforms. *International Journal of Contemporary Hospitality Management* 32(10), 3177–3197.
Nieuwland, S. and van Melik, R. (2020) Regulating Airbnb: how cities deal with perceived negative externalities of short-term rentals. *Current Issues in Tourism* 23(7), 811–825.
Nunnally, J.C. (1978) *Psychometric Theory*, 2nd edn. McGraw-Hill, New York.
Pike, A., Dawley, S. and Tomaney, J. (2010) Resilience, adaptation and adaptability. *Cambridge Journal of Regions, Economy and Society* 3(1), 59–70.
Ramos, S.P. and Mundet, L. (2020) Tourism-phobia in Barcelona: dismantling discursive strategies and power games in the construction of a sustainable tourist city. *Journal of Tourism and Cultural Change* 19(1), 113–131. https://doi.org/10.1080/14766825.2020.1752224
Ranchordas, S., Gedeon, Z. and Zurek, K. (2016) Home-sharing in the digital economy: the cases of Brussels, Stockholm, and Budapest. Impulse Paper prepared for the European Commission, DG GROW. Available at: https://ec.europa.eu/docsroom/documents/16950/attachments/1/translations/en/renditions/native (accessed 3 January 2021).
Richardson, L. (2015) Performing the sharing economy. *Geoforum* 67, 121–129.
Seravalli, A. (2014) Making commons (attempts at composing prospects in the opening of production). Doctoral thesis, Malmo University, Malmo, Sweden.
Sigala, M. (2017) The commercialization of the sharing economy: an investigation of the sub-economies of Airbnb. Paper presented at CAUTHE 2017 conference, 8 February 2017, University of Otago, Dunedin, New Zealand.
Slee, T. (2017) *What's Yours Is Mine: Against the Sharing Economy*. OR Books, New York.
Smorto, G. (2016) Impulse paper on the business authorisation/licensing requirements imposed on peer-providers and platforms in the accommodation/tourism sector in Paris, Rome, Milan and London. Impulse Paper NO.02. Available at: https://ec.europa.eu/docsroom/documents/16949/attachments/1/translations/en/renditions/native (accessed 3 January 2021).
Stergiou, D.P. and Farmaki, A. (2019) Resident perceptions of the impacts of P2P accommodation: implications for neighbourhoods. *International Journal of Hospitality Management* 91: 102411.
Thomas, K.A. and Clifford, S. (2017) Validity and mechanical turk: an assessment of exclusion methods and interactive experiments. *Computers in Human Behavior* 77, 184–197.
Van Doorn, N. (2020) A new institution on the block: on platform urbanism and Airbnb citizenship. *New Media & Society* 22(10), 1808–1826.
Volgger, M., Pforr, C. and Reiser, D. (2018) Destination resilience and the sharing economy: an exploration of the case of Airbnb. In: Innerhofer, E., Fontanari, M. and Pechlaner, H. (eds) *Destination Resilience: Chal-*

lenges and Opportunities for Destination Management and Governance. Routledge, London, pp. 142–152.
Wachsmuth, D. and Weisler, A. (2018) Airbnb and the rent gap: gentrification through the sharing economy. *Environment and Planning A: Economy and Space* 50(6), 1147–1170.

Wilson, G.A. (2012) Community resilience, globalization, and transitional pathways of decision-making. *Geoforum* 43(6), 1218–1231.

Appendix

Overview of 57 perceived impact items

Question prompt: To what extent, do short-term rental (STR) platforms (e.g. Airbnb) have an impact on your neighbourhood regarding the aspects listed below?

Impact item	Response[a]
The preservation of wildlife habitats	
The preservation of natural areas	
The preservation of cultural/historical sites	
Clean air and water	
Peace and quiet	
The beauty of my community	
Quality recreation opportunities	
Litter issues	
Traffic issues	
Crowding and congestion	
Drug and alcohol abuse	
Crime and vandalism	
Urban sprawl and population growth	
Conflicts over zoning/land use	
A feeling of belonging in my community	
The preservation of my way of life	
Authenticity of the neighbourhood	
Having tourists who respect my way of life	
Resident participation in local government	
My personal life quality	
Community pride	
Participation in local culture	
An understanding of different cultures	
Festivals, fairs, and museums	
The image of my community to others	
Awareness of natural and cultural heritage	
Having live sports to watch in my community	
Fair prices for goods and services	
City services like police and fire protection	
Good public transportation	
Offering of STRs in general	
Demand of STRs in general	
Hotel offering	
Hotel demand	
Attractiveness of the city	

Continued

Continued.

Impact item	Response[a]
Transport	
Attractions	
Income	
Employment	
Ethnic diversity	
Bohemian environment	
Melting pot environment	
Education level in the area	
Living conditions	
Green space	
House prices	
Property ownership by the local population	
Number of houses sold per square kilometre	
Overall price levels	
Working conditions in my city	
Employee rights in my city	
Hidden/black economy practices	
Regulation with regards to consumer protection, health and safety	
Environmental friendliness	
Regeneration of neighbourhoods in decline	
Rents	

[a]Response options: 1 – No impact; 2 – Very negative impact; 3 – Somewhat negative impact; 4 – Ambivalent impact; 5 – Somewhat positive impact; 6 – Very positive impact.

6 Airbnb Guests' Pro-environmental Behaviour and Community Resilience: Mitigating the Negative Impacts of Airbnb Tourism

Jinghua Xie* **and Olga Gjerald**
University of Stavanger, Norway

Abstract
Despite the significant amount of literature on tourists' pro-environmental behaviour (PEB), there is no study to systematically investigate the PEB of Airbnb guests. Understanding Airbnb guests' PEB is important to improve the host communities' resilience and mitigate the negative impacts associated with Airbnb growth. This chapter examines the determinants to Airbnb guests' PEB focusing on guests' knowledge of and concern about the environment. A binary logit model was applied using the survey data collected from Norwegian university students. The results support the hypothesis that Airbnb guests' environmental knowledge and concern are important in influencing their PEB. The study further tested if Airbnb guests behave differently when they are in or close to their home and when they travel. The findings show that respondents' behaviour does not seem to differ significantly. The study results enhance understanding of the Airbnb's impact on host communities and highlight the need to design PEB strategies specifically targeting Airbnb guests for successful destination governance and improved community resilience.

Keywords: Airbnb, pro-environmental behaviour, community resilience, sustainability, tourism, Norway

Highlights

- Understanding Airbnb guests' PEB is important to improve the host communities' resilience and mitigate the negative impacts associated with Airbnb growth.
- Airbnb guests' environmental knowledge has a significant positive influence on their PEB.
- Airbnb guests' environmental concern has a significant positive influence on their PEB.
- Airbnb guests have similar PEB intentions when they are staying in Airbnb and at home in the context of Norway.

Introduction

Prior to the COVID-19 outbreak, overtourism was identified as a major obstacle for achieving the goal of sustainable development of

*jinghua.xie@uis.no

 Peer-to-peer Accommodation and Community Resilience (eds Anna Farmaki, Dimitri Ioannides and Stella Kladou)

DOI: 10.1079/9781789246605.0006

destinations worldwide (UNWTO, 2017; Oklevik, *et al.*, 2019). Researchers have identified a wide range of causes of overtourism including: (i) the growing number of middle classes who have a high propensity to travel (UNWTO, 2017; Xie and Tveterås, 2020); (ii) falling travel costs (Ferrante *et al.*, 2018); and (iii) high seasonal concentration (Li *et al.*, 2018; Xie, 2020). Peer-to-peer (P2P) accommodation, mainly Airbnb rentals, has grown significantly in recent years and is blamed for being at least one of the causes of overtourism (Gössling *et al.*, 2020; Such-Devesa *et al.*, 2021). First, Airbnb rentals are frequently cheaper than conventional hotel accommodation making travel more affordable (Guttentag, 2015) and increasing tourism demand. Second, Airbnb penetrates residential neighbourhoods and while it allows visitors to 'live like a local', it also creates conflicts between visitors and residents because of problems emerging from the influx of tourists in neighbourhoods including noise, vandalism, safety, waste management and competition for public service (Gurran and Phibbs, 2017; Sklar and Edwards, 2017).

In this context, guests' pro-environmental behaviour (PEB) was highlighted as a significant factor influencing the impacts of Airbnb rentals on local communities. In the hotel setting, various actions have been put forward to encourage the PEB of guests (Kollmuss and Agyeman, 2002) including: (i) limiting car usage (Stanford, 2014); (ii) reusing towels and linen (Shang *et al.*, 2010); (iii) conserving water; (iv) reducing waste; and (v) choosing green food and beverages (Han *et al.*, 2018). Indeed, the PEB of hotel guests has been widely studied in the literature (Dolnicar *et al.*, 2017; Han, 2020) given the immense pressures placed on the environment and the local community as a result of hotels' large-scale operations and excessive waste (Farmaki and Stergiou, 2020). However, to the best of our knowledge, there is no systematic study on the Airbnb guests' PEB, despite the fact that Airbnb has grown tremendously over the last decade at a higher rate than the hotel sector (Cheng and Jin, 2019). This omission is surprising, as the rapid growth of the platform has brought about numerous problems for the local community of destinations, threatening its health and well-being (Stergiou and Farmaki, 2020).

Existing studies into the perceived impacts of Airbnb (Sklar and Edwards, 2017; Wegmann and Jiao, 2017; Cheng *et al.*, 2020b) and their relation to destination resilience (Volgger *et al.*, 2018) mostly explore economic and sociocultural aspects of sustainability. Among the positive impacts of Airbnb on destination resilience, several observers have documented increased competition, increased level of innovation, flexible inventory, diversification in markets and visitor segments, increased dispersal of income and spending, updating of regulatory frameworks, and positive host–guest encounters (Fang *et al.*, 2016; Tussyadiah and Pesonen, 2016; Volgger *et al.*, 2018). Perceived negative impacts of the Airbnb on destination resilience are elements of unfair competition, reduction of neighbourhood amenity and residents' quality of life, reduced level of safety and security, potential discrimination and community well-being issues, and negative impact on the relative return of tourism investment (Cheng and Foley, 2018; Kakar *et al.*, 2018; Volgger *et al.*, 2018). Meanwhile, few studies address the environmental impacts of Airbnb on destinations (Gurran and Phibbs, 2017; Cheng *et al.*, 2020a, b) especially the PEB of Airbnb guests.

The main objective of this chapter is to address an existing gap in the literature and examine the determinants of Airbnb guests' PEB, focusing on their knowledge of and concern about the environment. Since Airbnb has both positive and negative impacts on host communities (Cheng *et al.*, 2020b), it is important to take a closer look at Airbnb guests' PEB in the host community in order to understand the degree of the community's resilience against Airbnb penetration. Guests' PEB is important as it may help to reduce the negative impacts of Airbnb activity on the local community and provide knowledge which ultimately improves the local community's resilience. Drawing from a Norwegian setting, the study contributes significant theoretical and practical implications.

The rest of the chapter is organized as follows. It starts with a literature review of the topic of PEB and development of hypotheses. After that, we discuss the research context and data collection which are followed by a presentation of findings. We conclude with the main implications of the study.

Literature Review

In light of the growing problem of overtourism, research on tourists' PEB has drawn increasing attention since tourism's negative impacts are often associated with their delinquent behaviour in travel destinations (Pearce, 2005; Su and Swanson, 2017; Li and Wu, 2020). Most relevant studies measure tourists' PEB in tourist attractions such as national parks (Ramkissoon and Mavondo, 2014), ski resorts (Chen *et al.*, 2020) and zoos (Bueddefeld and Van Winkle, 2018). There is also a wealth of literature on green hotels (Millar *et al.*, 2012). Green hotels are hospitality products that actively practise environmentally responsible management activities (Line *et al.*, 2017; Han, 2020) in line with the growing importance attributed to corporate social responsibility in the sector (Farmaki, 2019). The guests of green hotels are often encouraged to actively practise PEBs such as reusing towels, conserving water and reducing waste (Dolnicar *et al.*, 2017; Line *et al.*, 2017) as well as pro-socially responsible behaviours including observing public order, being careful with the use of public facilities, showing courtesy to others, and taking part in healthy entertainment (Liu *et al.*, 2020).

A considerable number of studies have assessed the potential factors that may shape or facilitate responsible tourist behaviour. Most of these describe individual characteristics (Li and Wu, 2020) such as age, gender, education and income (Dolnicar, 2004; Hovardas and Poirazidis, 2006; Millar *et al.*, 2012; Patel *et al.*, 2017) or personality traits (e.g. openness to experience and agreeableness) and value including biospheric, socio-altruistic, or egoistic values (Markowitz *et al.*, 2012; Perkins and Brown, 2012). The importance of environmental knowledge and concerns in individuals' environmental attitudes and behaviours has long been documented in the literature (Schahn and Holzer, 1990; Stern *et al.*, 1993). In general, studies of PEB in travel and tourism show that people's pro-environmental values trigger PEBs (Ünal *et al.*, 2018). However, there are a few exceptions. For instance, Dolnicar *et al.*'s (2017) experimental study indicates that PEB appeals fail to significantly increase tourists' hotel towel reuse and decrease room electricity consumption, suggesting that interventions in hedonic contexts such as tourism may require the use of more tangible benefits in order to change behaviours.

Researchers have further recognized that while individuals are relatively comfortable with participating in a range of environmental behaviours in and around the home, the transference of these practices to tourism contexts is often problematic (Barr *et al.*, 2010; Dolnicar *et al.*, 2017). For instance, Barr *et al.* (2010) uncovered the fact that travellers with higher levels of commitment at home also tended to be those who flew furthest and most frequently, while those with moderate-to-high commitments of 'reducing, reusing, recycling' at home often failed to transfer these activities to their holiday environments. This means travellers might fail to transfer their PEB activities from their home environments to their holiday settings, indicating a gap between the theory and practice.

Despite the growing research interest in both tourists' PEB as well as the growth of Airbnb, there is a lack of studies systematically investigating PEB among Airbnb guests. To the best of our knowledge, this is the first study to empirically connect the concepts of PEB and Airbnb in relation to community resilience. Recent studies have explored the attitudes of 'green' Airbnb users (i.e. users associated with sustainable lifestyles) towards the accommodation they selected (Serrano *et al.*, 2021) as well as the impact of Airbnb on host communities (e.g. Oskam and Boswijk, 2016; Sklar and Edwards, 2017; Wegmann and Jiao, 2017; Cheng *et al.*, 2020a, b). Yet, most of them explore the economic and sociocultural impacts of the phenomenon. For instance, Oskam and Boswijk (2016) find that while Airbnb brings new income to locals, the increasing presence of Airbnb raises housing prices (Wegmann and Jiao, 2017) and rents (Horn and Merante, 2017) in the local host community.

Sklar and Edwards (2017) and Gurran and Phibbs (2017) report that Airbnb leads to increase in noise, parking and traffic congestion, and reduce the level of safety and security in neighbourhoods. This is in line with previous research on perceived tourism impacts on host communities (e.g. Gjerald, 2005). The most recent study by Cheng *et al.* (2020b) confirmed all of the above positive and negative effects in their examination of Airbnb-related impacts in Queenstown, New Zealand. The negative impacts identified in their study included: (i) increased cost of living; (ii) long-term housing shortage; (iii) high rental cost; (iv) increased housing prices and noise; (v) parking and traffic congestion; (vi) decreased safety; and (vii) waste management problems. Their research also identified diverse stakeholder perspectives on Airbnb impacts and how these perspectives led to different response strategies and anticipated outcomes.

In addition, Volgger *et al.*'s (2018) research of destination resilience and the sharing economy documented increased competition, increased level of innovation and flexible room inventory, diversification in markets and visitor segments in the hotel and tourism industry as well as updating of regulatory frameworks in host communities as significant factors.

Despite these important findings, very few studies address the environmental impact of Airbnb on destinations. Cheng *et al.* (2020a) assessed Airbnb's carbon footprint in Sydney and found that the growth of Airbnb had increased the carbon footprint. They argued that a significant amount of carbon emissions generated from induced consumption such as household energy has been ignored in the current environmental sustainability discourse around Airbnb. Furthermore, the extra income of Airbnb hosts can in turn potentially lead to social inequality and new consumption and production, inducing a carbon footprint. Moreover, Gurran and Phibbs (2017) documented that Airbnb visitors in Sydney, Australia are ignorant about local waste management. Similarly, Cheng *et al.* (2020b) reported that Airbnb guests generated additional waste for residential and commercial spaces, creating pressures for waste management.

Our study takes a broader look at various kinds of Airbnb guests' PEB and discusses the implications of these behaviours for community resilience. The object of this study is to examine Airbnb guests' knowledge and concerns about the environment. We argue that an Airbnb guest's PEB is fundamentally driven by his or her knowledge of and concern about the environment, as it is expected that those who have more knowledge/concern about environmental problems are more willing to take actions to protect the environment. Furthermore, as opposed to a hotel guest, an Airbnb guest often stays in rentals in residential areas (Stergiou and Farmaki, 2020). Although Airbnb supply has the flexibility to locate themselves close to the tourism attraction and/or established hotels (Eugenio-Martin *et al.*, 2019), considering the number of hosts with multiple listings was 23.5% in January 2019 (Shatford, 2019), it means most of the Airbnb rooms are shared by residents living in neighbourhoods. Therefore, this study's aim is to examine whether the finding given by Barr *et al.* (2010), that individuals are relatively more willing to participate in environmental behaviours in and around the home than in a travel environment, also applies to Airbnb tourists. We, thus, tested the following three hypotheses:

- Hypothesis 1: Airbnb guests' environmental knowledge has a significant positive influence on their PEB.
- Hypothesis 2: Airbnb guests' environmental concern has a significant positive influence on their PEB.
- Hypothesis 3: Airbnb guests have lower PEB intentions when they are staying in Airbnb than at home.

Research Context and Questionnaire Formulation

Norwegian Airbnb guests were chosen as the context for this research. As in other countries around the world, Airbnb has experienced a significant annual growth in Norway. According to the hotel data provided by Statistics Norway (2020) and Airbnb data provided by AirDNA (2020), the number of Airbnb rooms booked increased by 138%, 45% and 28% in 2017, 2018 and 2019, respectively. By contrast, the growth rates for hotel overnight stays were modest (2–5% yearly) for the same period. Demand for Airbnb rentals has grown faster than demand for hotel rooms, which makes Airbnb a significant contributor to tourism growth in Norway.

Using data for the period between January 2016 and August 2020, we find the average prices were 900–1000 Norwegian krone (NOK) for hotel rooms and 400–500 NOK for Airbnb rooms, resulting in an average difference of 536 NOK between the two forms of accommodation. We also witnessed that both Airbnb and hotels have peak season between June and September, suggesting the lower price of Airbnb has attracted more tourists in peak seasons and consequently accelerated seasonal overtourism in Norway.

For this study, we used a survey to identify the Airbnb guests' PEB in host communities around Norway. The survey questionnaire was written in English and comprised four sections. The first section included social-demographic data relating to age, gender, education, income and other variables to identify the respondents' travel purposes and length of stay. The second section measured visitors' PEB. Specifically, it included the popular environmentally friendly activities of recycling paper and cardboard, reusing plastic bags, using grocery bags, and buying organic groceries following Millar *et al.* (2012) and buying local products following Bieak (2010). Since the PEBs of hotel guests in hotel rooms, such as lowering the room temperature and reusing towels, have been widely studied in the green hotel literature, we chose to include similar questions relating to these matters, subject to some adjustments. These questions were designed following the literature on green hotels (Gustin and Weaver, 1996; Kreidler, 2010; Millar and Baloglu, 2011; Han *et al.*, 2018) and were adjusted with reference to Barr *et al.* (2010), who compared the individuals' PEB in and around the home and while travelling. Finally, sections three and four included environmental knowledge and concerns, respectively, by using measurement instruments by Millar *et al.* (2012) and Cottrell (2003). We classified environmental concerns into two separate categories: (i) 'optimistic concern'; and (ii) 'pessimistic concern'. Specifically, the questions showing people are positive about the future environment were categorized as positive concerns while those showing that they are negative or worried about the future environment were treated as pessimistic concerns.

Table 6.1 presents the detailed PEB questions while Table 6.2 shows measurements of environmental knowledge and concerns. All items were designed along a Likert scale (1 = lowest, 7 = highest) to measure the extent to which respondents differed in their evaluations of specific PEBs, their knowledge and concern about the environment. Tables 6.1 and 6.2 summarize the literature that inspired our selection of items.

An online survey was conducted using a non-probability convenience sampling method. The survey was sent out to all the students at the authors' home institute using their university email addresses. To enhance variation in responses, the questionnaire was also sent to students' social media groups. Of the total 293 questionnaires we have received, 198 had no missing variables and were, therefore, used in the study. Due to the COVID-19 pandemic, the

Table 6.1. Airbnb guests' pro-environmental behaviour (PEB) in host communities.

Behaviour	References
Recycle paper and cardboard	Millar *et al.* (2012)
Reuse plastic bag	
Use cloth grocery bag	
Buy organic food	
Use public transportation	Developed by the authors
Buy local products	Bieak (2010)
Lower the temperature when leaving Airbnb rental	Adapted from Gustin and Weaver (1996) and Millar and Baloglu (2011)
Turn off lights when leaving Airbnb rental	
Comparison with behaviour at home	
Wash the clothing more often in Airbnb rentals than at home	Adapted from Gustin and Weaver (1996), Kreidler (2010), Millar and Baloglu (2011) and Han *et al.* (2018)
Have showers more often or use more water in Airbnb rentals than at home	
Wash towels and bedding more often in Airbnb rentals than at home	
Use tumble dryer to dry clothes more often in Airbnb rentals than at home	By the authors of the chapter
Use tumble dryer to dry towels and bedding more often in Airbnb rentals than at home	

Table 6.2. Airbnb guests' knowledge and concern about the environment.

Topics included in knowledge and concern	References
Knowledge	Cottrell (2003), Millar *et al.* (2012)
Informed about water pollution	
Informed about the effects that treatment of raw sewage has on the quality of the marine environment	
Informed about the local environmental problems and issues where they live	
Knowledgeable about ecology	
Knowledgeable about environmental issues related to energy	
Knowledgeable about environmental issues related to CO_2 emissions	
Knowledgeable about how environmental issues are related to waste	
Concern (pessimistic)	Cottrell (2003), Millar *et al.* (2012)
We are approaching the limit of the number of people the Earth can support	
Despite our special abilities, humans are still subject to the laws of nature	
Humans are severely abusing the environment	
If things continue on their present course, we will soon experience a major ecological catastrophe	
Concern (optimistic)	
The so-called 'ecological crisis' facing humankind has been greatly exaggerated	
The balance of nature is strong enough to cope with the impacts of modern industrial nations	
Human ingenuity will ensure that we do *not* make the Earth unliveable	
Earth has plenty of natural resources if we just learn how to develop them	

sample size is relatively small but contributes to a first assessment of the research hypotheses.

Using a student sample may lead to the problem of generalizability of findings. Therefore, we call for caution when generalizing the findings of this study to a larger population of Airbnb guests. However, despite the limitation, the results of our study provide initial insights into the PEB of young travellers and suggest directions for future research. Moreover, young people are the most important market segment for Airbnb. As suggested by both literature (Guttentag, 2015) and statistics (Statistics Norway, 2019), young, technologically comfortable, adventurous and budget-conscious tourists present the main characteristics of Airbnb guests. Furthermore, young consumers represent the future consumer so their knowledge, concern, attitude and behaviours to protect the environment provides a better understanding of the direction in which we are headed (Meinhold and Malkus, 2005).

Methodology

The study used the econometric model of the binary logit model to identify the factors affecting Airbnb visitors' PEB in host communities. The dependent variable in the model is PEB_i, a specific PEB such as recycling paper and cardboard, lowering the room temperature and so on.

$$PEB_i = \begin{cases} 1 \, if \; an \; AirBnB \; visitor \; showed \; PEB_i \\ 0, \; else \end{cases} \qquad (6.1)$$

The probability of an Airbnb guest shows PEB_i is a function of the factors influencing this PEB intention.

$$Prob[PEB_i = 1] = \frac{e^{x'\beta}}{1+e^{x'\beta}} \quad (6.2)$$

The empirical specification of $z = x'\beta$ in the study is:

$$Z_{PEB_i} = \alpha_{i0} + \alpha_{i1}Knowledge + \alpha_{i2}Pessimisticconcern + \alpha_{i3}Optimisticconcern + \alpha_{i4}Gender + \alpha_{i5}Age + \alpha_{i6}Education + \alpha_{i7}Income + \varepsilon_i \quad i = 1, 2, 3, \ldots 8 \quad (6.3)$$

where the explanatory variables are factors influencing Airbnb guests' PEB. These refer to Airbnb guests' knowledge and concerns about the environment and social-demographic variables of gender, age, education and income as indicated by the variable names. Concern has been broken into pessimistic concern and optimistic concern to identify the different behaviours of pessimists and optimists as discussed above. As presented in Table 6.2, knowledge is measured by seven items (questions), pessimistic concern by four items and optimistic items by another four items. i denotes each PEB investigated in the study. There are altogether eight PEBs as presented in the estimated results (see Table 6.3 in the next section) and also in the first part of Table 6.1.

At this point, we evaluated two options. Opting for the first option suggests undertaking a factor analysis which then generates composite variables (Mehmetoglu, 2014; Xie *et al.*, 2020). However, the factor analysis results showed that all the items were loaded in the proper factors and that the loadings in each factor were quite close. Therefore, our study employs the second option and used average weights instead. This approach is more straightforward and provides similar results compared to using factor loadings as weights. As a result, we have first generated a composite variable for each factor (i.e. knowledge, pessimistic concern and optimistic concern) by taking the means of the respondents' scores for the items in each factor. Preliminary analysis also found that travel purpose and length of stay are not statistically significant in any of the PEB behaviour models. Thus, they were consequently both excluded from the final estimated model.

If denoting $x'\beta$ as z, and the above cumulative probability function as $F(z)$, Eqn 6.2 can be rewritten as:

$$F(z) = \frac{e^{z}}{1+e^{z}} \quad (6.4)$$

The marginal effect of any explanatory variable x_i on the probability of $PEB_i = 1$ (the estimated results presented in Table 6.3) can be derived by:

Table 6.3. Estimated results of the models.[a]

	Dependent variable						
PEB	Knowledge	Pessimistic concern	Optimistic concern	Gender	Age	Education	Income
Recycle paper and cardboard	–	–	–	13%	–	–	–
Reuse plastic bag	–	11%	–	–	–	–	–
Use cloth grocery bag	15%	–	–	–18%	–	11%	–5%
Buy organic food	14%	–	–	–12%	9%	–	–5%
Use public transportation	–	21%	–	–	–	–	–4%
Buy local products	–	19%	–	–	13%	–	–
Lower the temperature when leaving Airbnb facility	18%	–	–	–	7%	–	–
Turn off lights when leaving Airbnb facility	–	20%	–	–	–	–	–

[a]Only the estimated coefficients that are statistically significant at a 10% critical level or lower are reported; otherwise, they are noted as '–'.

$$\frac{dF}{dx_i} = \beta_i F(z).(1-F(z)) \quad (6.5)$$

An ordered logit model should be more appropriate as PEB is measured by a seven-point Likert scale ordered from 1 = lowest extent to 7 = highest extent. However, we face the same common issue discussed by Mehmetoglu (2014) and Xie *et al.* (2020): this is that the data is highly skewed with few respondents scoring 3 or lower. Therefore, based on the above literature we aggregated the respondents into two categories. For a specific PEB *i* (e.g. recycling paper and cardboard), those who have scored 5 or higher are the group who have strong intentions to show PEB *i* and the rest who have scored 4 or lower are the lower intention group. This makes a binary logit model that was finally estimated in the study. In the binary model, the dependent variable PEB_i is a binary choice decoded as a dummy. As presented by Eqn 6.1, it is 1 if an Airbnb visitor has a strong intention to show PEB_i, otherwise 0 (zero).

Results

As Table 6.3 illustrates, knowledge and concerns are proven to be important in influencing Airbnb guests' PEB. Knowledge is important in influencing guests' behaviour when using cloth grocery bags, buying organic food, and lowering the temperature when leaving Airbnb facilities. Specifically, for Airbnb guests who have better knowledge about the environment there is a 15% greater possibility of their using a cloth grocery bag than those who have less knowledge. The corresponding enhanced possibilities are 14% and 18%, respectively, for buying organic food and lowering the temperature when leaving Airbnb facilities. Environmental concern is also found to be important. The positive results of pessimistic concern suggest that those who are pessimistic about the future environment take action to protect the environment. High environmental concern increases the possibility of Airbnb guests reusing plastic bags, using public transportation, purchasing local products and turning off lights by 11%, 21%, 19% and 20%, respectively. No estimated parameter of optimistic concern is found to be statistically significant in any PEB model, suggesting that those with optimistic concern do not have a strong PEB intention. The results indicate that people are rational: although those who have optimistic concerns have a lower intention to protect the environment, they will nevertheless not harm the environment. The results of concern are, therefore, as expected.

Among the social-demographic variables of gender, age, education and income, we found education to be least important since the variable is only significant in the equation of using cloth grocery bags. One reason for this is because we used a student sample, the education variation in the achieved sample was small as most of them were bachelor students. Compared to education, variation in income is relatively high. However, the negative estimated coefficients of the income variable are not expected since it means those who have higher incomes have a lower intention to behave in an environmentally friendly manner. The age results suggest the older people are, the more willing they are to exhibit PEB. This is somewhat expected since several studies linking PEB and age (e.g. Patel *et al.*, 2017) show that a more mature demographic (age group 36–50 years old) is more likely to display PEB orientation compared with the younger adults (age group 20–35 years old and younger). Although some researchers have pointed out that environmentally friendly consumers are generally younger (Getzner and Grabner-Kraüter, 2004), others report that increase in environmentally friendly behaviour and awareness is positively associated with increase in age (Tilikidou, 2007; Abeliotis *et al.*, 2010).

The gender results show men are 13% more likely than women to recycle paper and cardboard. Conversely, women are 18% more inclined than men to use cloth grocery bags and purchase organic food. These results are in line with previous research where males and females display differential PEBs (Xiao and Hong, 2010; Chen *et al.*, 2011).

The third hypothesis investigates whether Airbnb guests have lower PEB intentions when they are staying in Airbnb accommodation than when they are at home. In order to address this hypothesis, we focused on the survey results deriving from the questions listed in the second part of Table 6.1 ('Comparison with behaviour at home'). To make the results more readable, we recoded answers using a three-point scale: aggregating scores from 1 to 3 feeds into the 'mostly disagree' category, scores from 5 to 7 feeds into the 'mostly agree' category, while 4 is 'neutral'. Our results show most of the respondents did not behave differently when at home as opposed to an Airbnb rental (Table 6.4). For example, 69% of the respondents answered they did not wash the clothing more often in Airbnb, 66% answered they did not have showers more often in Airbnb, and 73% answered they would not wash towels and bedding more often in an Airbnb property than they would at home.

Discussion and Conclusion

This study contributes to extant literature by investigating the Airbnb guests' PEB, which is important for community resilience. Guests' PEB will mitigate the negative impact of a growing number of tourists on the local community. Understanding the Airbnb guests' PEB is, therefore, important for communities' strategies to improve their resilience against the phenomenon. Based on the analysis of the knowledge and concern variables, we accept the first and second hypotheses that Airbnb guests' knowledge and concerns have significant positive influences on their PEB in host communities. The third hypothesis that individuals have lower PEB intentions when they are away from home is not supported, since the dominant share of the respondents disagree that they behave differently in an Airbnb rental from how they behave at home. It should be mentioned, however, that the sample is not representative of the entire population of travellers since the sample included students, and they may vary from other groups of travellers (i.e. older travellers). Future research should test this hypothesis using non-student samples.

The results relating to knowledge and concern are consistent with most of the findings in the literature on PEB (Schahn and Holzer, 1990; Stern *et al.*, 1993; Cottrell, 2003; Millar *et al.*, 2012). However, the result of the third hypothesis is inconsistent with the finding given by Barr *et al.* (2010) that individuals struggle to transfer their environmental behaviour practices in and around the home to a tourism context. The reason for this might be that Airbnb guests stay in Airbnb rentals which resemble a home, unlike hotel settings. Practically, Airbnb guests might continue with their household routines as psychologically they might not feel that a 'holiday is a holiday' (Barr *et al.*, 2010) aiming to relax rather than upkeep the PEB. In hotel settings, there are employees who are responsible for waste management and recycling, whereas in Airbnb units it is the guests who are frequently in charge of such tasks (depending on the length of stay, type of property, etc.). Therefore, we argue that compared with hotel guests, Airbnb guests are more responsible for everyday environmental practices such as recycling and reusing materials, purchasing local products and using public transportation. More research is needed on how Airbnb guests' PEBs differ from hotel guests' PEBs as well as residents in neighbourhoods where Airbnb rentals are located.

Although there have been complaints about Airbnb guests' ignorance of local waste

Table 6.4. PEB of Airbnb guests at home and in travel.

Behaviour	Mostly disagree (%)	Neutral (%)	Mostly agree (%)
I wash the clothing more often in Airbnb rentals than at home.	69	17	14
I have showers more often or use more water in Airbnb rentals than at home.	66	19	15
I wash towels and bedding more often in Airbnb rentals than at home.	73	16	11
I use a tumble dryer to dry clothes more often in Airbnb rentals than at home.	68	16	16
I use a tumble dryer to dry towels and bedding more often in Airbnb rentals than at home.	68	19	13

management (Gurran and Phibbs, 2017), we did not find evidence that Airbnb guests are less responsible environmentally. It seems they behave like residents when they choose to experience local living in an Airbnb rental. Nevertheless, comparative studies in other countries are needed since Norwegians in general are environmentally friendly and have good knowledge and are highly concerned about the environment. This might be not true in other countries. Furthermore, there is a lack of infrastructure with regard to recycling and waste management in many countries, which in turn limits guests' ability to carry out responsible, pro-environmental acts. Besides, there is a need to examine the host's role in encouraging guests to behave in a more environmentally friendly manner. In a broader perspective, this resonates with a research call to investigate new forms of sustainable hospitality experiencescapes (Gjerald *et al.*, 2021), where the P2P accommodation sector is a significant contributor.

In relation to community resilience and Airbnb growth, this study provides knowledge that can be applied to mitigate the negative impacts of Airbnb growth in host communities. We argue that it is important for successful destination governance to design PEB strategies targeting Airbnb guests. For instance, local communities and Airbnb hosts can provide information to Airbnb guests about how they can contribute to lower the negative impacts of Airbnb and save the local community's resources. The information about local policies for promoting sustainable practices in destinations such as saving energy and purchasing local products should also be made available to Airbnb guests visiting the destination. Local stakeholders can look at green hotel practices (Kreidler, 2010; Millar *et al.*, 2012) when designing policies for Airbnb hosts (e.g. using occupancy sensors and natural lights, leaving messages and reminders in Airbnb rooms) to inform guests about the environmental problems and encourage them to demonstrate specific PEB. Similar to promoting 'green hotels' in the hotel industry, local communities can encourage Airbnb hosts to promote 'green Airbnb' to attract and retain more environmentally friendly Airbnb visitors to stay in the communities.

A key aspect of resilience is the process of fostering a degree of diversity that enables communities to be both robust and flexible at the same time (Hartman, 2016). A degree of diversity would mean that visitors may be reached by PEB strategies regardless of where they stay – hotels or Airbnb rentals. In this way, local PEB initiatives will not fail when particular accommodation providers are used. Resilience is destination specific and thus there is no one-size-fits-all approach that will provide the right mix between regulation and innovation for all contexts (Dredge and Gyimóthy, 2017; Hall *et al.*, 2017). The Airbnb phenomenon and the destination governance responses will influence each other in the years to come, and community resilience will be affected by adaptive destination governance responses to Airbnb growth.

References

Abeliotis, K., Koniari, C. and Sardianou, E. (2010) The profile of the green consumer in Greece. *International Journal of Consumer Studies* 34, 153–160.

AirDNA (2020) AirDNA Data: How it Works. Available at: https://www.airdna.co/airdna-data-how-it-works#:~:text=%20AirDNA%20Data%3A%20How%20it%20Works%20%201,is%20core%20to%20any%20analysis%20of...%20More%20 (accessed 20 October 2020).

Barr, S., Shaw, G., Coles, T. and Prillwitz, J. (2010) 'A holiday is a holiday': practicing sustainability, home and away. *Journal of Transport Geography* 18(3), 474–481. DOI: 10.1016/j.jtrangeo.2009.08.007

Bieak, K.N. (2010) An examination of green environmental attributes in hotel service settings. PhD dissertation, Auburn Alabama University, Alabama.

Bueddefeld, J.N. and Van Winkle, C.M. (2018) The role of post-visit action resources in facilitating meaningful free-choice learning after a zoo visit. *Environmental Education Research* 24(1), 97–110.

Chen, F., Dai, S., Zhu, Y. and Xu, H. (2020) Will concerns for ski tourism promote pro environmental behaviour? An implication of protection motivation theory. *International Journal of Tourism Research* 22(3), 303–313.

Chen, X., Peterson, M.N., Hull, V., Lu, C., Lee, G.D., *et al.* (2011) Effects of attitudinal and sociodemographic factors on pro-environmental behaviour in urban China. *Environmental Conservation* 38, 45–52.

Cheng, M. and Foley, C. (2018) The sharing economy and digital discrimination: the case of Airbnb. *International Journal of Hospitality Management* 70, 95–98. DOI: 10.1016/j.ijhm.2017.11.002

Cheng, M. and Jin, X. (2019) What do Airbnb users care about? An analysis of online review comments. *International Journal of Hospitality Management* 76, 58–70. DOI: 10.1016/j.ijhm.2018.04.004

Cheng, M., Chen, G., Wiedmann, T., Hadjikakou, M., Xu, L. and Wang, Y. (2020a) The sharing economy and sustainability – assessing Airbnb's direct, indirect and induced carbon footprint in Sydney. *Journal of Sustainable Tourism* 28(8), 1083–1099.

Cheng, M., Houge Mackenzie, S. and Degarege, G.A. (2020b) Airbnb impacts on host communities in a tourism destination: an exploratory study of stakeholder perspectives in Queenstown, New Zealand. *Journal of Sustainable Tourism* July, 1–19. DOI: 10.1080/09669582.2020.1802469

Cottrell, S.P. (2003) Influence of sociodemographics and environmental attitudes on general responsible environmental behavior among recreational boaters. *Environment and Behavior* 35(3), 347–375.

Dolnicar, S. (2004) Insights into sustainable tourists in Austria: a data-based a priori segmentation approach. *Journal of Sustainable Tourism* 12(3), 209–218. DOI: 10.1080/09669580408667234

Dolnicar, S., Knezevic Cvelbar, L. and Grün, B. (2017) Do pro-environmental appeals trigger pro-environmental behavior in hotel guests? *Journal of Travel Research* 56(8), 988–997. DOI: 10.1177/0047287516678089

Dredge, D. and Gyimóthy, S. (eds) (2017) *Collaborative Economy and Tourism*. Springer, Berlin.

Eugenio-Martin, J.L., Cazorla-Artiles, J.M. and González-Martel, C. (2019) On the determinants of Airbnb location and its spatial distribution. *Tourism Economics* 25(8), 1224–1244.

Fang, B., Ye, Q. and Law, R. (2016) Effect of sharing economy on tourism industry employment. *Annals of Tourism Research* 57, 264–267. DOI: 10.1016/j.annals.2015.11.018

Farmaki, A. (2019) Corporate social responsibility in hotels: a stakeholder approach. *International Journal of Contemporary Hospitality Management* 31(6), 2297–2320.

Farmaki, A. and Stergiou, D.P. (2020) Corporate social responsibility and employee moral identity: a practice-based approach. *Current Issues in Tourism* 24(18), 2554–2572. DOI: 10.1080/13683500.2020.1850654

Ferrante, M., Magno, G.L.L. and De Cantis, S. (2018) Measuring tourism seasonality across European countries. *Tourism Management* 68, 220–235.

Getzner, M. and Grabner-Kraüter, S. (2004) Consumer preferences and marketing strategies for green shares. *The International Journal of Bank Marketing* 22, 260–278.

Gjerald, O. (2005) Sociocultural impacts of tourism: a case study from Norway. *Journal of Tourism and Cultural Change* 3(1), 36–58.

Gjerald, O., Dagsland, Å.H.B. and Furunes, T. (2021) 20 years of Nordic hospitality research: a review and future research agenda. *Scandinavian Journal of Hospitality and Tourism* 21 (1), 1–12.

Gössling, S., Scott, D. and Hall, C.M. (2020) Pandemics, tourism and global change: a rapid assessment of COVID-19. *Journal of Sustainable Tourism* 29, 1–20, DOI: 10.1080/09669582.2020.1758708

Gurran, N. and Phibbs, P. (2017) When tourists move in: how should urban planners respond to Airbnb? *Journal of the American Planning Association* 83(1), 80–92.

Gustin, M.E. and Weaver, P.A. (1996) Are hotels prepared for the environmental consumer? *Hospitality Research Journal* 20(2), 1–14.

Guttentag, D. (2015) Airbnb: disruptive innovation and the rise of an informal tourism accommodation sector. *Current Issues in Tourism* 18(12), 1192–1217.

Hall, C.M., Prayag, G. and Amore, A. (2017) *Tourism and Resilience: Individual, Organizational and Destination Perspectives*. Channel View Publications, Bristol, UK.

Han, H. (2020) Theory of green purchase behavior (TGPB): a new theory for sustainable consumption of green hotel and green restaurant products. *Business Strategy and the Environment* 29(6), 2815–2828. DOI: 10.1002/bse.2545

Han, H., Lee, J.-S., Trang, H.L.T. and Kim, W. (2018) Water conservation and waste reduction management for increasing guest loyalty and green hotel practices. *International Journal of Hospitality Management* 75, 58–66.

Hartman, S. (2016) Towards adaptive tourism areas? A complexity perspective to examine the conditions for adaptive capacity. *Journal of Sustainable Tourism* 24(2), 299–314. DOI: 10.1080/09669582.2015.1062017

Horn, K. and Merante, M. (2017) Is home sharing driving up rents? Evidence from Airbnb in Boston. *Journal of Housing Economics* 38, 14–24.

Hovardas, T. and Poirazidis, K. (2006) Evaluation of the environmentalist dimension of ecotourism at the Dadia Forest Reserve (Greece). *Environmental Management* 38(5), 810–822.

Kakar, V., Voelz, J., Wu, J. and Franco, J. (2018) The visible host: does race guide Airbnb rental rates in San Francisco? *Journal of Housing Economics* 40, 25–40.

Kollmuss, A. and Agyeman, J. (2002) Mind the Gap: why do people act environmentally and what are the barriers to pro-environmental behavior? *Environmental Education Research* 8(3), 239–260. DOI: 10.1080/13504620220145401

Kreidler, N.B. (2010) An examination of green environmental attributes in hotel service settings. Doctoral thesis, Auburn University, Auburn, Alabama. Available at: https://etd.auburn.edu/handle/10415/2334 (accessed 11 January 2021).

Li, H., Goh, C., Hung, K. and Chen, J.L. (2018) Relative climate index and its effect on seasonal tourism demand. *Journal of Travel Research* 57(2), 178–192.

Li, Q. and Wu, M. (2020) Tourists' pro-environmental behaviour in travel destinations: benchmarking the power of social interaction and individual attitude. *Journal of Sustainable Tourism* 28(9), 1371–1389. DOI: 10.1080/09669582.2020.1737091

Line, N.D., Hanks, L. and Miao, L. (2017) Image matters: incentivizing green tourism behavior. *Journal of Travel Research* 57(3), 296–309.

Liu, J., An, K. and Jang, S. (2020) A model of tourists' civilized behaviors: toward sustainable coastal tourism in China. *Journal of Destination Marketing & Management* 16: 100437. DOI: 10.1016/j.jdmm.2020.100437

Markowitz, E.M., Goldberg, L.R., Ashton, M.C. and Lee, K. (2012) Profiling the 'pro-environmental individual': a personality perspective. *Journal of Personality* 80(1), 81–111. DOI: 10.1111/j.1467-6494.2011.00721

Mehmetoglu, M. (2014) Predictors of sustainable consumption in a tourism context: a CHAID approach. In: Chen, J.S. (ed.) *Advances in Hospitality and Leisure*. Emerald Group Publishing, Bingley, UK, pp. 3–23.

Meinhold, J.L. and Malkus, A.J. (2005) Adolescent environmental behaviors: can knowledge, attitudes, and self-efficacy make a difference? *Environment and Behavior* 37(4), 511–532.

Millar, M. and Baloglu, S. (2011) Hotel guests' preferences for green guest room attributes. *Cornell Hospitality Quarterly* 52(3), 302–311.

Millar, M., Mayer, K.M. and Baloglu, S. (2012) Importance of green hotel attributes to business and leisure travelers. *Journal of Hospitality Marketing & Management* 21(4), 395–413.

Oklevik, O., Gössling, S., Hall, C.M., Steen Jacobsen, J.K., Grøtte, I.P. and McCabe, S. (2019) Overtourism, optimisation, and destination performance indicators: a case study of activities in Fjord Norway. *Journal of Sustainable Tourism* 27(12), 1804–1824.

Oskam, J. and Boswijk, A. (2016) Airbnb: the future of networked hospitality businesses. *Journal of Tourism Futures* 2(1), 22–42.

Patel, J., Modi, A. and Paul, J. (2017) Pro-environmental behavior and socio-demographic factors in an emerging market. *Asian Journal of Business Ethics* 6(2), 189–214.

Pearce, P.L. (2005) *Tourist Behaviour: Themes and Conceptual Schemes*. Aspects of Tourism series. Channel View Publications, Clevedon, UK. DOI: 10.21832/9781845410247

Perkins, H.E. and Brown, P.R. (2012) Environmental values and the so-called true ecotourist. *Journal of Travel Research* 51(6), 793–803. DOI: 10.1177/0047287512451133

Ramkissoon, H. and Mavondo, F. (2014) Pro-environmental behavior: the link between place attachment and place satisfaction. *Tourism Analysis* 19(6), 673–688.

Schahn, J. and Holzer, E. (1990) Studies of individual environmental concern: the role of knowledge, gender, and background variables. *Environment and Behavior* 22(6), 767–786.

Serrano, L., Ariza-Montes, A., Nader, M., Sianes, A. and Law, R. (2021) Exploring preferences and sustainable attitudes of Airbnb green users in the review comments and ratings: a text mining approach. *Journal of Sustainable Tourism* 29(7), 1134–1152. DOI: 10.1080/09669582.2020.1838529

Shang, J., Basil, D. and Wymer, W. (2010) Using social marketing to enhance hotel reuse programs. *Journal of Business Research* 63(2), 166–172. DOI: 10.1016/j.jbusres.2009.02.012

Shatford, S. (2019) The short-term rental market is evolving. So is AirDNA. AirDNA, 4 March. Available at: https://www.airdna.co/blog/short-term-rental-market-evolving-homeaway (accessed 5 February 2021).

Sklar, W.P. and Edwards, J.C. (2017) Florida community associations versus Airbnb and VRBO in Florida. *Florida Bar Journal* 91(2), 16.

Stanford, D.J. (2014) Reducing visitor car use in a protected area: a market segmentation approach to achieving behaviour change. *Journal of Sustainable Tourism* 22(4), 666–683. DOI: 10.1080/09669582.2013.847944

Statistics Norway (2019) Bruk av internett til å ordne innkvartering fra andre privatpersoner. Statistikkbanken (ssb.no). Available at: https://www.ssb.no/statbank/table/12755 (accessed 12 January 2021).

Statistics Norway (2020) Transport and tourism – SSB. Available at: https://www.ssb.no/en/transport-og-reiseliv (accessed 3 March 2021).

Stergiou, D.P. and Farmaki, A. (2020) Resident perceptions of the impacts of P2P accommodation: implications for neighbourhoods. *International Journal of Hospitality Management* 91: 102411. DOI: 10.1016/j.ijhm.2019.102411

Stern, P.C., Dietz, T. and Kalof, L. (1993) Value orientations, gender, and environmental concern. *Environment and Behavior* 25(5), 322–348.

Su, L. and Swanson, S.R. (2017) The effect of destination social responsibility on tourist environmentally responsible behavior. *Tourism Management* 60, 308–321. DOI: 10.1016/j.tourman.2016.12.011

Such-Devesa, M.J., Ramón-Rodríguez, A., Aranda-Cuéllar, P. and Cabrera, A. (2021) Airbnb and overtourism: an approach to a social sustainable model using big data. In: Balsalobre-Lorente, D., Driha, O.M. and Shahbaz, M. (eds) *Strategies in Sustainable Tourism, Economic Growth and Clean Energy*. Springer, Cham, Switzerland, pp. 211–233.

Tilikidou, I. (2007) The effects of knowledge and attitudes upon Greeks' pro-environmental purchasing behaviour. *Corporate Social Responsibility and Environmental Management* 14, 121–134.

Tussyadiah, L. and Pesonen, J. (2016) Impacts of peer-to-peer accommodation use on travel patterns. *Journal of Travel Research* 55(8), 1022–1040.

Ünal, A.B., Steg, L. and Gorsira, M. (2018) Values versus environmental knowledge as triggers of a process of activation of personal norms for eco-driving. *Environment and Behavior* 50(10), 1092–1118. DOI: 10.1177/0013916517728991

UNWTO (World Tourism Organization) (2017) *Yearbook of Tourism Statistics, Data 2011–2015*, 2017 edition. UNWTO, Madrid. DOI: 10.18111/9789284418428

Volgger, M., Pforr, C. and Reiser, D. (2018) Destination resilience and the sharing economy: an exploration of the case of Airbnb. In: Innerhofer, E., Fontanari, M. and Pechlaner, H. (eds) *Destination Resilience: Challenges and Opportunities for Destination Management and Governance*. Routledge, Abingdon, UK, pp. 142–152.

Wegmann, J. and Jiao, J. (2017) Taming Airbnb: toward guiding principles for local regulation of urban vacation rentals based on empirical results from five US cities. *Land Use Policy* 69, 494–501.

Xiao, C. and Hong, D. (2010) Gender differences in environmental behaviours in China. *Population and Environment* 32, 88–104.

Xie, J. (2020) The economic determinants of tourism seasonality: a case study of the Norwegian tourism industry. *Cogent Business & Management* 7(1): 1732111.

Xie, J. and Tveterås, S. (2020) Economic decline and the birth of a tourist nation. *Scandinavian Journal of Hospitality and Tourism* 20(1), 49–67.

Xie, J., Tkaczynski, A. and Prebensen, N.K. (2020) Human value co-creation behavior in tourism: insight from an Australian whale watching experience. *Tourism Management Perspectives* 35: 100709.

7 Reframing Rurality: the Impact of Airbnb on Second-home Communities in Wales and Sweden

Bailey Ashton Adie,[1,2]* Cecilia de Bernardi[3,4] and Alberto Amore[1]
[1]Solent University, UK; [2]Wakayama University, Japan; [3]Dalarna University, Sweden; [4]Mid-Sweden University, Sweden

Abstract

This chapter presents evidence from two established second-home destinations (Snowdonia in the UK and the Dalarna region in Sweden), where the proliferation of properties listed on Airbnb is currently contributing to the reframing of the existing rural landscape. This study suggests that there is a separate but connected layer of spatial interaction inherent in these rural environments, which is mediated by the perceptions of rural space of the three user groups: (i) primary residents; (ii) second homeowners; and (iii) Airbnb users. This chapter argues that the influence of peer-to-peer (P2P) platforms within these mixed-use communities injects an additional level of spatial interaction and conceptualization of rurality. Based on the findings from this study, it is noted that Airbnb guests' perceptions of rurality are similar to those of other rural tourists, but this is also influenced by the local hosts and wider rural community. Understanding how these users perceive the rural space is essential to the development of resilient rural communities as consensus is critical for resilience planning.

Keywords: second homes, P2P, rurality, rural tourism, rural landscape, rural communities, netnography

Highlights

- Airbnb reviews are analysed through a qualitative netnography.
- Different dimensions of rurality between Airbnb guests and other local residents are discussed.
- The netnography clearly shows important aspects for Airbnb users such as proximity to amenities and attractions.
- Understanding these perspectives is essential to the development of resilient rural communities.

Introduction

Rural areas have long been popular tourist destinations. Second homes, in particular, are often found in these locations, and, in Europe, this phenomenon is closely tied to the imagery of the rural in opposition to the urban, a common theme within rural studies (Bell, 2006; Cloke, 2006). In fact, rurality, according to Kamvasinou and Stringer (2019, p. 784), 'is promoted through the marketing of second homes to an urban elite'. While the second-home tradition, particularly in Scandinavia, is often tied to traditional

*bailey.adie@gmail.com

DOI: 10.1079/9781789246605.0007

seasonal utilitarian population movements from the city to rural areas, recent second-home acquisitions are frequently the result of shifting populations, particularly from rural to urban areas (Müller, 2007). However, although second homeowners and primary residents have often coexisted in these spaces for some time, the recent rise in sharing platforms like Airbnb have added another layer of social complexity to these rural spaces. Research on the rise of Airbnb and similar peer-to-peer (P2P) platforms is gaining momentum, yet it predominantly focuses on urban or mature coastal destinations. To date, research on these platforms in rural areas is limited, and, in particular, the impact of sharing-platform users on countryside second-home communities remains unstudied. The need for research into sharing platforms and second homes has been noted in the literature (Marjavaara *et al.*, 2019) as has the call for a better understanding of resilience and tourism in rural areas (Helgadóttir and Dashper, 2021). What emerges is the need for clarity and evidence with regards to resilience as a social construct resulting from the iteration of tourism-relevant stakeholders at the destination level (Amore *et al.*, 2018). It is these noted gaps that our research seeks to fill. In particular, the emphasis on the rural allows for a more focused analysis without the spatial and population complexities inherent in urban environments. In order to provide more depth to the study, we selected two case studies where Airbnb is at different stages of development, one in Sweden and one in Wales.

Theory

Traditionally, rurality, as a theoretical construct, has been, and continues to be in many areas of rural studies, associated with the dichotomy between physical urban and rural environments (Bell, 2006; Cloke, 2006; Holden and Lupton, 2017). However, as Cloke (2006, p. 19) notes, a firm distinction between these two spaces no longer exists, leading to their conceptualization of rurality as a social construct 'characterized by a multiplicity of social spaces overlapping the same geographical area'. This then allows for multi-layered understanding and interpretations of what constitutes an area's 'rurality' as defined by the various rural space users (Woods, 2010), which Crouch (2006, p. 363) refers to as 'a kaleidoscope of culturally mediated images'. Rurality is often framed as an act of consumption as opposed to production (Jepson and Sharpley, 2015) although the heterogeneity of interpretation of this rural space by different users can cause conflict (DuPois, 2006). For example, rural tourists often search for a sanitized rural idyll that is embedded in a nostalgic interpretation of the rural past (Heldt Cassel and Pettersson, 2015; Panzer-Krause, 2020) or even simply conceived as a space devoid of human activity entirely (Bell, 2006; Holden and Lupton, 2017). This can often conflict with the lived rural realities of local residents. As a case in point, according to research by Heldt Cassel and Pettersson (2015), when local farmers in Sweden developed rural farm-based tourist experiences, it was necessary to present an idealized farm environment, which was both sterile and tranquil. This was meant to appeal to their visitors' sense of what constitutes an authentic rural environment, in contrast with the modern farms as sites of production.

However, it would be incorrect to separate groups so succinctly into locals and non-locals as both rurality and social groupings are exceedingly more complex. Even among permanent residents, there can be contested perceptions of rural space. This is clearly visible in cases of significant in-migration, as was observed in Smith and Phillips' (2001) study of the gentrification process of a village in Yorkshire, UK. They identified two specific groups whose views on rurality, while both rooted in an idealized rural past, were divergent depending on where they had chosen to live in the area. Village dwellers were more focused on the rural idyllic village lifestyle, and the local community in particular, whereas those who chose to purchase housing on the moors were more concerned with solitude and pristine space. Both of these views of rurality appear similar to those observed among rural tourists, but, in the case of permanent residents, the emphasis on community resulted in the integration and assimilation of the village in-migrants, easing their inclusion into the local space. In contrast, the moor-based in-migrants, similar to tourists and predominantly due to their rejection of local community and customs, found their understanding of local rurality at odds with more traditional rural culture (Smith and

Phillips, 2001). Thus, within residential subgroups, contrasting images of rurality may exist, but the levels of acceptance by different group members are dependent on community involvement.

The importance of community is a common theme in the discussion of rural spaces, and it is highly relevant, in the case of this chapter, for the discussion of second homes in these rural spaces. While the phrase 'second home' has no agreed-upon universal definition (Hall and Müller, 2004; McIntyre, 2006; Paris, 2009; Müller, 2014), it is often used as a blanket term to allow for a variety of contexts and usage (Hall and Müller, 2004; Back and Marjavaara, 2017). In rural areas, second homes are predominantly either purpose built or purchased, although some are inherited family homesteads (McIntyre *et al.*, 2006; Williams and Van Patten, 2006; Flemsæter, 2009). In all instances, second homeowners are spatially integrated with local permanent residents and may be reliant on similar amenities, such as grocery stores, hardware stores and cafés (Larsson and Müller, 2019; Tuulentie and Kietäväinen, 2020). However, similar to rural tourists and the in-migrants in the study by Smith and Phillips (2001), second homeowners' interpretation of the rurality of the space in which their properties are located is framed either as pristine wilderness (Vepsäläinen and Pitkänen, 2010) or through the nostalgic rural idyll, where, in contrast to tourists who are often shielded from the realities of rural life, there is an active avoidance of non-idyllic rural elements (Pitkänen, 2008). None the less, rural second homeowners, much like many of their primary counterparts, place significant value on the local community and their situatedness in it (Vepsäläinen and Pitkänen, 2010; Nordin and Marjavaara, 2012), even when this community may only be composed of other second homeowners (Tuulentie and Kietäväinen, 2020). Thus, second homeowners who are more attached to the community, similar to the majority of primary residents, will mediate their spatial understanding through their interpersonal relationships.

In this context, the rise of Airbnb and similar P2P accommodation platforms for temporary lettings in rural areas contributes to an additional layer of complexity in the social construct of rurality. Research on Airbnb in rural areas is at an early stage, with evidence from Italy and Spain showing a higher ratio of listings per capita in the idyllic countryside of Tuscany and Sierra Nevada (Cesarani and Nechita, 2017; Adamiak *et al.*, 2019). However, the focus of most existing studies is on the distribution of listed properties in the countryside (e.g. Cors-Iglesias *et al.*, 2020). Additionally, studies on the perception of Airbnb in rural areas are limited and fall short in providing a clear appraisal of Airbnb's role in host–guest relationships and community well-being (Gyimóthy and Meged, 2016; Suess *et al.*, 2021). There is nevertheless research demonstrating how escapism, localness and authenticity are generally relevant when choosing between a listed property on Airbnb and other types of accommodation (Mody *et al.*, 2017; Guttentag *et al.*, 2018). These studies, however, do not delve into the imagery of specific localities and tend to focus on the shortcomings of the hospitality sector from an experience-economy perspective.

Given the growth of these P2P properties and their position within local communities, there is a need to better understand these users from a holistic perspective in order to enhance community resilience, which is the ability of individuals and groups to withstand short- and long-term shocks to their social infrastructure (Adger, 2000; Norris *et al.*, 2008; Lew, 2014). Studies highlight the proactive role of the community in destination resilience building (Lawton and Weaver, 2015; Sheppard and Williams, 2015), but restrict their research to the views of local residents. The importance of technological innovations in relation to community resilience in rural areas has been noted, and research in these areas is gaining momentum. Evidence from the UK shows how rural digital inclusion is central in rural community empowerment and resilience from a socio-technological perspective (e.g. Roberts *et al.*, 2017). The nexus between technology and resilience enables rural communities 'to take responsibility and control of their rural development pathways' (Wilson, 2010, p. 366), and thus transform work life and job markets in the foreseeable future (Ashmore *et al.*, 2017). From a community resilience perspective, Airbnb and other short-term letting platforms represent an emerging yet fragmented research focus. There is acknowledgement of the importance of 'investigating the benefits but also the costs of such activities in local communities' (Ioannides, 2016, p. 436), but the phenomenon of second homes has been overlooked. Similarly, there is an awareness of the proactive role second homeowners can play from both a destination resilience and adaptability perspective (Hall, 2016a; Amore *et al.*, 2018), but there is no explicit mention of Airbnb and similar P2P platforms.

This dearth of research is problematic when considering the importance of understanding how different space users conceptualize their environment and the values that they attribute to it (Amore *et al.*, 2018). In rural areas with second homes, Airbnb users tend to inhabit the same spaces as primary and secondary homeowners and consume the same services. Nevertheless, it can be assumed, due to their limited temporal involvement with the rural area, that their perceptions of rurality are more similar to those previously attributed to rural tourists in general. These Airbnb users then may ascribe different values to different local assets through their understanding of the area's rurality. Thus, they exist in a liminal rural space, being neither homeowners nor temporary rural consumers. This is an issue that has yet to be addressed in the literature.

Methodology

Context and data collection

In order to understand how Airbnb users understand rurality within mixed second-homeowner and primary-resident communities, we selected two case studies, namely Älvdalen municipality in Sweden and Conwy and Gwynedd in Wales. Both of these locations are rural areas with pre-existing second-home communities as well as a growing number of available Airbnb rentals, albeit with the Swedish growth in Airbnb being much more recent. For the Swedish case, we selected the area of Älvdalen municipality, which also includes Idre, Idrefjäll, Fulufjället National Park and the areas of Älvdalen in the south (see Table 7.1). Idre and Älvdalen are located in the Dalarna region in mid-Sweden. Both are close to Norway and have traditionally been frequented by second homeowners as well as tourists. Visitors to the Dalarna region are predominantly domestic, with the next largest source markets being the other Nordic countries, the Netherlands and Germany (SCB and Tillväxtverket, 2019). This area is also a popular alpine skiing destination (Berglund, 2005) although only approximately 7000 people permanently reside there (Älvdalens kommun, 2019). The area under study consisted of approximately 110 active rentals. Data were collected between July and October 2020, resulting in a total of 1867 analysed reviews. These were predominantly in Swedish, but there were also reviews in other Scandinavian languages, German, Dutch and English.

The case study municipalities in the UK are more heavily populated than those in the Swedish context. Therefore, to ensure sound comparability between the two cases, we examined properties in municipalities which are close to a set of main natural points of interest (Table 7.2), in particular those with listed properties in the vicinity of Snowdonia National Park Visitor Centre, main natural attractions, and train stations. Two municipalities, Conwy and Gwynedd, met these criteria. The proximity to the Midlands and the North-West of England means that Conwy and Gwynedd are a desired rural location

Table 7.1. Main points of interest in Älvdalen municipality, Sweden.

Point of interest	Location (number of properties analysed)
National park	Storsätern (9), Björnliden (6), Storbo (3), Görddalen (6), Mörkret (5), Heden (3), Särna (3), Fulufjället National Park (1), Nornäs (1)
Skiing facilities	Höstsätern (2), Idrefjäll (25), Fjätervålen (4)
Body of water	Foskros (4), Drevdagen (1), Lövnäs (1), Åsen (1), Rämma (2), Karlsarvet (3), Brunnsberg (1), Näset (1)
Central/close to major road	Älvdalen (4), Idre (11)
More rural isolated area	Evertsberg (8)
Nature reserve	Blyberg (5)

Table 7.2. Main points of interest in selected municipalities in Snowdonia, Wales.

Point of interest	Municipality	Location (number of properties analysed)
National park	Conwy	Glanaber Terrace/Cwm Penmachno (5), Ysgol Latymer (3), Carrog (4), Gwydir Forest Park (1), Penmacho (27), Snowdonia National Park (41)
	Gwynedd	Beddgelert (25)
Body of water	Conwy	Capel Curig (8)
	Gwynedd	Trawsfynydd (13), Bala (15)
Train station	Gwynedd	Dolwyddelan (17)
More rural isolated area	Conwy	Ysbyty Ifan (10)

for temporary lettings among domestic tourists and underpins the existing data for average length of stay available before the rise of Airbnb in the UK (ONS, 2020). Between the two municipalities, 169 listed properties were analysed resulting in a total of 4671 reviews being collected between July and October 2020. Unlike the Swedish data, the reviews were predominantly in English, with the remainder being in French, German and Spanish.

Data analysis

We derived our empirical data from a netnographic content analysis of Airbnb and Vrbo as visualized on the website AirDNA.co (n.d.), which compiles data from both websites and displays it as a map. Netnography is a method that is used to investigate people's interactions and postings online (Mkono and Markwell, 2014), but there is great variation in how the method is used (Mkono and Markwell, 2014). In this chapter, we undertook a content analysis to assess the data (Elo *et al.*, 2014). The reviews were collected and translated if the language was not one of those spoken by the researchers. In order to analyse the data, the reviews were imported into an Excel file where different elements of the reviews were coded into different columns representing identified categories. An inductive coding process was used in order to ensure that the coding was representative of the data. To this end, one researcher was in charge of the analysis (Elo *et al.*, 2014), while the others focused on the theoretical development of the topic. This ensured that the data analysis was interpreted through the literature but not biased by it. The reviews were categorized in terms of: (i) the location of the P2P properties as defined by AirDNA.co (n.d.); (ii) the number of stars given by the guests indicating their level of satisfaction; (iii) the time range of the reviews for a certain property (from oldest to newest); and (iv) whether or not a host reply was present. The time range indicated provisionally if a property has been listed for a long time or not.

The data was sorted into four categories that described the content. These were: (i) *facilities* (see Table 7.3); (ii) *location* (see Table 7.4); (iii) *nature/rurality* (see Table 7.5); and (iv) *host* (see Table 7.6). Comments which fell outside of these classifications were placed into a category labelled *other*, but the majority were later redistributed to the other categories as no new categories emerged. This is a standard procedure in both content and thematic analyses (Elo *et al.*, 2014; Walters, 2016). The reviews were assigned keywords, commonly known as codes, which were reworked as we collected more data. The final categories are those previously identified, and they reflect different aspects of the Airbnb experience, which were deemed worthy of note.

Findings

Sweden

In Sweden, the results of the netnography illustrate several Airbnb rental review trends. The first of these relates to the *facilities* category (Table 7.3), which includes elements related directly to the rented accommodation. These descriptions tended to focus on the interior attributes of the properties, such as kitchens, bathrooms and bedrooms, with an emphasis on their utilitarian aspects. This emphasis on practical elements was also observed in relation to the second category,

Table 7.3. Examples of facility-related reviews.

Case study	Review extracts
Sweden	'The modern kitchen is wonderful, with plenty of utensils, pots and pans and dishes'
	'Beds galore, multiple kitchens, bathrooms and showers'
	'Cabin was adequately furnished'
Wales	'Very homely cottage'
	'The cottage is lovely and atmospheric'
	'Well equipped and cosy rustic farmhouse'
	'Very homely and cosy cottage'

Table 7.4. Examples of location-related reviews.

Case study	Review extracts
Sweden	[The property is] 'easy to find and close to hiking trails'
	[The property is] 'close to a number of skiing slopes'
	'The National park Fulufjäll is 15 min away by car'
	'Very cosy cabin with a perfect location [läge] to the Njuperskärs waterfall!'
	[The property is] 'comfortable and close to everything but anyway isolated [avskilt]'
	'The cottage only had two other cottages nearby and the rest was beautiful forest Nonetheless we were still within walking distance of a shop […]'
	'Close to convenience store and village centre'
	'Vicinity to shops and food places'
Wales	'It is a short walk into town if you'd like a drink at dinner'
	'Close to village shops'
	'Excellent location in a village with a shop and a pub'
	'No need to drive anywhere, cafés and restaurants just a short walk away'
	'Location is ideal with a great walk behind the cottage'
	'Really well-placed for walking and access to all the adventurous activities'
	'Located in a charming town right down the road from a Snowden Peak trail head and many more to boot'
	'Great location for a stroll along the river to Beddgelert'
	'Location is perfect, leafy and close to stunning walks'
	'Very good location for visiting Snowdonia National Park'
	'The location of the cottage is excellent!! In the heart of the beautiful Betws Y Coed'
	'Just a short walk from the centre of picturesque Betws Y Coed'

location (Table 7.4). There were several reviews which specifically highlighted the importance of the properties' proximity to different amenities, such as shops and convenience stores. In addition to the functional needs of the Airbnb users, the reviews in this category also underscored the properties' vicinity to key points of local interest, such as ski slopes, the national park, or a waterfall. Furthermore, guests were appreciative of properties which were not only close to the aforementioned points of local interest but also retained a sense of isolation. Therefore, these reviews are illustrative of visitor preferences for both convenient access to amenities and activities while retaining a sense of solitude.

While reviews in the *location* category included proximity to specific natural features, the *nature/rurality* category (Table 7.5) included vicinity to nature more broadly as well as more general observations related to the overall rural setting of the property. Many guests highly regarded the natural setting of the cabins, with several juxtaposing the rural with the urban, which highlights this continued view of the countryside as being in opposition to the city. Additionally, these natural surroundings were often viewed through the guest's own sociocultural

Table 7.5. Examples of nature/rurality-related reviews.

Case study	Review extracts
Sweden	'Comfortable and cosy accommodation close to nature'
	'The location is fantastic with nature [...] "on the doorstep"'
	'The house is in the proper Swedish wilderness next to the national park [...]'
	'The Stuga is perfect for relaxing and enjoying pure Swedish nature'
	'Exotically located cabin in a quiet and silent snow-filled forest [...]'
	'We are a couple who wanted to get away from the city life and experience nature and the feeling of living in a cabin in the woods [...]'
	'Lovely charming house far away from city life and noise'
	[The cabin is placed in] 'an idyllic forest environment'
	[The cabin is] 'in wonderful nature'
	[The cabin is] 'with nice nature around'
	Nature described as 'untouched'
	A cabin described as 'surrounded by fantastic hosting [værtskab] and nature'
	'It is definitely a magical place!'
	A guest appreciates the 'closeness to fantastic nature and magical mountains'
	A property described as a 'typically Swedish cottage [...]'
	A property described as a 'small cottage typical in Sweden [...]'
	Guests 'saw more reindeer than cars on that street'
	There are 'many reindeer and some moose'
	A property in which there are chickens clucking described 'like out of a picture book [...]'
Wales	'Cute traditional stone cottage in a tiny village'
	'Lodge set in beautiful surroundings'
	'Romantic village idyll'
	'Romantic'
	'The cottage offered that level of quietness and isolation that we wanted'
	'It is pretty isolated (apart from a few surrounding houses) but that's exactly what we were after'
	'Chance to really get away from it all'
	'Absolutely idyllic location with stunning views'
	'The surrounding countryside and location are fabulous'
	'The cottage is a quiet valley, and the scenery was beautiful'
	'Awesome house in an idyllic location, immaculate and a great place to stay the weekend'
	'Beautiful view of Snowdon in the morning'
	'Spectacular views, history and countryside'
	'Such a beautiful village surrounded by jaw dropping scenery'
	'Stunning sunsets over the lake'
	'A beautiful valley view and so many cute animals'

lens. This is particularly evident with non-Nordic foreign reviewers who emphasized the typically 'Swedish' elements of the natural environment. For example, one reviewer referred to their rental as 'a charming little traditional Swedish cabin with a warm homey feel [...].' Furthermore, and less common in both foreign and domestic reviews, there is also mention of the rural as idyllic. While the word 'idyllic' is only used in a handful of reviews, there are other, much more common synonyms used, such as 'picturesque', 'romantic', 'magical' and 'authentic'. The inclusion of domestic farm animals as an important element of the rural experience can also be understood as part of this idyllic focus, with the presence of chickens clucking at one property being described as something 'out of a picture book [...].' The final focus of reviews in this category is on the rural space as wilderness, with one review referring to it as 'untouched'. This included mention of wild animals, such as moose and reindeer.

The final category, *host* (Table 7.6), was a less recurring topic within the reviews and was concentrated on the actual Airbnb host–guest interactions. These comments stressed the friendliness and

Table 7.6. Examples of host-related reviews.

Case study	Review extracts
Sweden	'Host was very responsive to messages [...]'
	'The host was really nice and communicative :)'
	'The host was friendly and responsive'
	'We got tips and guide for places to visit [...]'
	The host is 'a great guide to the region[,] knows everything about the animals, the history, what's going on'
	The 'host does not live at the place [...]'
	The guests and the owner 'did not meet live [...]'
	The owners tell that they 'aren't normally present [them]selves [...]'
	'The cabin is located right next to the one [the host] uses'
	'The cottages they offer are perfectly located'
Wales	'Hosts are very welcoming'
	'Hosts could not have been more hospitable'
	'Host great in terms of communication and recommendations'
	'Recommendations for places to eat and her advice for climbing Snowdon was also brilliant'
	'Great tips for local walks'
	'We appreciated the local recommendations from host'
	'Host very helpful and kept us very well informed and gave great advice on things to do on the local area'

communicativeness of the hosts while also underscoring the importance of the hosts' breadth of local knowledge. One guest noted that the Airbnb host was 'a great guide to the region[,] knows everything about the animals, the history, what's going on.' In some instances, the hosts were multiple property owners with a second home for personal use in the area. However, this is not to say that all hosts were present, and, in certain cases, the host was not a resident of the area at all. Thus, local host interaction with guests was not guaranteed, and, in these cases, local interaction may be assumed to have been limited given the lack of any reviews discussing locals.

In summary, Airbnb guests in the Swedish case are particularly interested in a rural experience which is rooted in solitude within a natural setting. In particular, there is an emphasis on the isolation of both properties and individuals. Given the relative newness of Airbnb in this setting, this could potentially lead to problems in the future wherein new properties built for guests will also destroy this very element of isolation that rural Airbnb users seek.

Wales

The data from Wales, as in Sweden, was separated into the identified four categories: (i) *facilities*; (ii) *location*; (iii) *nature/rurality*; and (iv) *hosts*. In the Welsh context, the comments in the *facilities* (Table 7.3) category differed substantially from the Swedish findings in that they were more likely to be affective as opposed to utilitarian. Cottages were often referred to as 'homely' or 'rustic', emphasizing comfort and a rural aesthetic. The results from the reviews, which appear in the *location* category (Table 7.4), however, are more similar to those found in Sweden. More specifically, there is a significant emphasis placed on the property's proximity to attractions like Snowdonia National Park as well as local activities, such as walking and bike trails as well as outdoor activities. Interestingly, and something that was absent in the Swedish case, are reviews noting the proximity to a local village, which one reviewer described as 'picturesque'. Finally, and again similar to Sweden, additional comments in this category related specifically to local businesses, especially shops and restaurants. The reviews stress that there are 'plenty of eateries at the local town' and how close most of these establishments are to the rental properties.

In the *nature/rurality* category (Table 7.5), reviews tend to emphasize the rustic nature of the listed property and the villages in the surrounding area. Comments referred, in particular, to the traditional elements of the listed cottages and the 'romantic village idyll'. This is interesting in comparison with the Swedish context where the rural was very much viewed as an experience apart from human activity. This is not to say that remoteness is not a quality noted by reviewers in Wales. In fact, one reviewer stated that 'the cottage offered that level of quietness and isolation that [the user] wanted'. There are further descriptions reinforcing the narrative of quietness and peacefulness of the site. For example, one review describes the area as 'peace and quiet and stunning scenery', while another refers to the area as an 'absolutely idyllic location with stunning views, offering the chance to really get away from it all'. Thus, in Wales, rurality was framed both as quiet wilderness as well as historic idyllic village.

The final category, *host* (Table 7.6), was again similar to the Swedish case. In particular, hosts were highlighted as being very 'hospitable' and 'welcoming' with the hosts being noted as the point of contact for the rest of the community. The reviews acknowledge the role that hosts play in providing information and advice on where to go and what to do while at the property. Visitors value their recommendations and tips on walks and outdoor activities as well as where to go for dinner. One review in particular illustrates how the guests 'felt very welcome and loved the little touches such as recommendations for local pubs/restaurants and important Welsh phrases'. The latter is of interest as some guests appear to consider local culture and knowledge as important elements of the Welsh rural experience. Given the importance of local culture, as well as the previously noted emphasis on villages as points of interest, it is perhaps unsurprising that reviews in Wales explicitly mention the local population, an element that was absent in the Swedish reviews. For example, one review refers to the locals as 'lovely friendly people full of good cheer'.

In summary, the reviews of the listed properties in Conwy and Gwynedd suggest that Airbnb guests seek a mix of remoteness from densely populated areas and proximity to picturesque cottages and villages. The reviews also suggest a preference for properties in tranquil and quiet areas surrounded by nature and an idealized Welsh rural landscape.

Discussion

Although the Swedish and Welsh cases are noticeably different in terms of the national origin of their key market as well as their level of Airbnb development, the findings for both have some marked similarities. In line with previous research on rural tourists (Bell, 2006; Holden and Lupton, 2017), the Airbnb guests in both locations stress the remoteness and quietness of the Airbnb locations, with their conceptualization of the local space's rurality emphasizing it as untouched wilderness. Additionally, the perception of the local environment as a rural idyll appears in both contexts, albeit only among international guests in Sweden. In the case of Wales, it can be argued that the cultural elements appearing in the *host* category enhance this sense of a Welsh rural idyll wherein local traditional elements and the Welsh language itself impact upon guests' view of the area's rurality. Additionally, guests in both Sweden and Wales highlight the importance of the location of local amenities in proximity to their rented property. While understandably many of these reviews are specifically referring to ease of access to local attractions and leisure activities, there is also a focus on the availability of shops and eateries, spaces which are also of importance to both local residents and second homeowners. Thus, there is a similar level of spatial integration between these guests and the other rural homeowners, both primary and secondary, within the public sphere (Larsson and Müller, 2019; Tuulentie and Kietäväinen, 2020).

Integration with the local population is further enhanced by the host–guest relationship apparent in the Airbnb sharing model. While rural tourists often engage with the local population through service-based interactions, Airbnb users are more likely to highlight interpersonal communication with the local hosts, or lack thereof in the case of Sweden. However, when the hosts do interact with the Swedish guests, they are often members of the local second-home community and perceived as fonts of local knowledge. This emphasis on the hosts as sources of information was also visible in Wales, but, in this context, there was additional reference to the wider local community and the guests' engagement with them. Therefore, the hosts, as well as the wider community in the case of Wales, play an integral role in the Airbnb guests' understanding of the local area. These important interpersonal, yet shallow, interactions are distinct from the deeper community connections, which influence second homeowners' and primary residents' perceptions of the local rural space. In the case of the Airbnb guests, they are external actors who engage with, but are distinct from, the local sociocultural environment. Consequently, their perceptions of local rurality may share elements from both primary and second homeowners while the lens through which they view the space remains exogenous.

The findings from this comparative study shed light on actor-related patterns in relation to the construct of rurality among the local community, second homeowners, and Airbnb guests in Älvdalen and Snowdonia. The importance of individual behaviours and perceptions is acknowledged in emerging research on destination resilience (Amore *et al.*, 2018), and, as was noted by DuPois (2006) heterogenous perceptions of rurality can cause conflict at the local level. Therefore, it is necessary to fully understand this new local actor group, Airbnb users, and how they can contribute to the strengthening of community resilience against potential internal and external shocks. However, from a socio-technological perspective, the Airbnb online community has the potential to enhance the resilience of rural communities. Building on recommendations by Hall (2016a), the inclusion of Airbnb guests can assist in the development of inclusive destination-level planning. For example, Airbnb can act as a catalyst towards more inclusive and community-driven modes of destination metagovernance. Arguably, these new mechanisms of decision making can empower second homeowners, rural residents and Airbnb guests as invested stakeholders and, in turn, contribute towards rural community resilience. These initiatives can spark regime shifts and, in turn, reframe the local conceptualization of rurality to be inclusive of multiple viewpoints. Additionally, it can potentially provide a stronger grassroots level of support in order to both adapt and respond to short- and long-term ecological triggers (Hall, 2016b).

Conclusions

Through an analysis of Airbnb reviews from Sweden and Wales, in this chapter we have sought to explore guests' understanding of local rural spaces, which are also popular second-home destinations. While it was noted that rural Airbnb guests exhibited similar views of rurality to those expressed by more general rural tourists, they seem to place an emphasis on amenities and interpersonal interactions with the local community. However, due to the brevity of these community interactions, the Airbnb guests are still external actors albeit with their perceptions of rurality partially moderated by the local community's viewpoints. From a community resilience perspective, it is essential to include all relevant viewpoints in order to 'build [...] consensus around destination resilience planning' (Amore *et al.*, 2018, p. 245). Therefore, it is important that those responsible for local-community resilience planning understand these P2P users better in order to ensure that future strategies are inclusive of the breadth of communities, both permanent and temporary, that occupy rural space at any given moment.

It should be noted that this chapter has several limitations. In particular, we only collected data on Airbnb users. Future research needs to also engage with the local permanent-resident and second-home communities in order to better understand the differences in spatial perceptions. Furthermore, rurality is a complex phenomenon, while in this chapter we have offered a predominantly generalist viewpoint. Future work needs to also look at the intersections of rurality in these rural spaces with race, gender, class, and so on. This is especially important given the potential demographic differences between primary homeowners, second homeowners and Airbnb guests. In particular, future research needs to provide a better understanding of Airbnb users in these spaces given their growing presence in rural locations.

References

Adamiak, C., Szyda, B., Dubownik, A. and García-Álvarez, D. (2019) Airbnb offer in Spain – spatial analysis of the pattern and determinants of its distribution. *International Journal of Geo-Information* 8, 155.

Adger, W.N. (2000) Social and ecological resilience: are they related? *Progress in Human Geography* 24, 347–364.
AirDNA.co. (n.d.). Short-term rental analytics I Vrbo and Airbnb Data. Available at: https://www.airdna.co/ (accessed 5 February 2021).
Älvdalens kommun (2019) Befolkning. Available at: https://www.alvdalen.se/kommun-och-politik/kommun-fakta/befolkning.html (accessed 6 November 2020).
Amore, A., Prayag, G. and Hall, C.M. (2018) Conceptualizing destination resilience from a multilevel perspective. *Tourism Review International* 22, 235–250.
Ashmore, F.H., Farrington, J.H. and Skerratt, S. (2017) Community-led broadband in rural digital infrastructure development: implications for resilience. *Journal of Rural Studies* 54, 408–425.
Back, A. and Marjavaara, R. (2017) Mapping an invisible population: the uneven geography of second-home tourism. *Tourism Geographies* 19(4), 595–611.
Bell, D. (2006) Variations on the rural idyll. In: Cloke, P., Marsden, T. and Mooney P. (eds) *The Handbook of Rural Studies*. SAGE Publications, London, pp. 149–160.
Berglund, C. (2005) Ja, vi elsker dette landet... og: En studie kring norskägda fritidshus i svenska delen av Inre Skandinavien. *Karlstads universitet, CERUT, arbetsrapporter* 15.
Cesarani, M. and Nechita, F. (2017) Tourism and the sharing economy. An evidence from Airbnb usage in Italy and Romania. *Symphonya. Emerging Issues in Management* 3, 32–47.
Cloke, P. (2006) Conceptualizing rurality. In: Cloke, P., Marsden, T. and Mooney P. (eds) *The Handbook of Rural Studies*. SAGE Publications, London, pp. 18–28.
Cors-Iglesias, M., Gómez-Martín, M.B. and Armesto-López, X.A. (2020) Peer-to-peer accommodation in rural areas of Catalonia: defining typologies of rural municipalities. *Sustainabilty* 12, 6145.
Crouch, D. (2006) Tourism, consumption and rurality. In: Cloke, P., Marsden, T. and Mooney P. (eds) *The Handbook of Rural Studies*. SAGE Publications, London, pp. 355–364.
DuPois, E.M. (2006) Landscapes of desires? In: Cloke, P., Marsden, T. and Mooney P. (eds) *The Handbook of Rural Studies*. SAGE Publications, London, pp. 124–132.
Elo, S., Kääriäinen, M., Kanste, O., Pölkki, T., Utriainen, K. and Kyngäs, H. (2014) Qualitative content analysis: a focus on trustworthiness. *SAGE Open January–March 2014*, 1–10.
Flemsæter, F. (2009) From 'home' to 'second home': emotional dilemmas on Norwegian smallholdings. *Scandinavian Journal of Hospitality and Tourism* 9(4), 406–423.
Guttentag, D., Smith, S., Potwarka, L. and Havitz, M. (2018) Why tourists choose Airbnb: a motivation-based segmentation study. *Journal of Travel Research* 57(3), 279–295.
Gyimóthy, S. and Meged, J.W. (2016) The shareable countryside: rescaling host–guest relationships in the collaborative economy. Paper presented at the 2016 RSA Workshop: Beyond the Great Beauty: Rescaling Heritage and Tourism, 10–12 February 2016, Universitá degli Studi di Bologna, Rimini, Italy.
Hall, C.M. (2016a) Intervening in academic interventions: framing social marketing's potential for successful sustainable tourism behavioural change. *Journal of Sustainable Tourism* 24(3), 350–375.
Hall, C.M. (2016b) Putting ecological thinking back into disaster ecology and responses to natural disasters: rethinking resilience or business as usual? In: Hall, C.M., Malinen, S., Vosslamber, R. and Wordsworth, R. (eds) *Post-disaster Management: Business, Organizational and Consumer Resilience and the Christchurch Earthquakes*. Routledge, Abingdon, UK, pp. 269–92.
Hall, C.M. and Müller, D.K. (2004) Introduction: second homes, curse or blessing? Revisited. In: Hall, C.M. and Müller, D.K. (eds) *Tourism, Mobility and Second Homes: Between Elite Landscape and Common Ground*. Channel View Publications, Bristol, UK, pp. 3–14.
Heldt Cassel, S. and Pettersson, K. (2015) Performing gender and rurality in Swedish farm tourism. *Scandinavian Journal of Hospitality and Tourism* 15(1–2), 138–151.
Helgadóttir, G. and Dashper, K. (2021) 20 years of Nordic rural tourism research: a review and future research agenda. *Scandinavian Journal of Hospitality and Tourism* 21(1), 60–69.
Holden, A. and Lupton, K. (2017) Experiencing and connecting to nature: an urban to rural association. In: Slocum, S.L. and Kline, C. (eds) *Linking Urban and Rural Tourism: Strategies in Sustainability*. CABI, Wallingford, UK, pp. 20–32.
Ioannides, D. (2016) Airbnb and the shared economy in Sweden: building resilience or illusions? In: Blàzquez, M., Mir-Gual, M., Murray, I. and Pons, G.X. (eds) *Turismo y Crisis, Turismo Colaborativo y Ecoturismo: XV Coloquio de Geografía del Turismo, el Ocio y la Recreación de la AGE*. Societat d'Història Natural de les Balears, Palma de Mallorca, Spain, pp. 436–437.

Jepson, D. and Sharpley, R. (2015) More than sense of place? Exploring the emotional dimension of rural tourism experiences. *Journal of Sustainable Tourism* 23(8–9), 1157–1178.
Kamvasinou, K. and Stringer, B. (2019) The politics of rurality. *Landscape Research* 44(7), 783–786.
Larsson, L. and Müller, D.K. (2019) Coping with second home tourism: responses and strategies of private and public service providers in western Sweden. *Current Issues in Tourism* 22(16), 1958–1974.
Lawton, L.J. and Weaver, D.B. (2015) Using residents' perceptions research to inform planning and management for sustainable tourism: a study of the Gold Coast Schoolies Week, a contentious tourism event. *Journal of Sustainable Tourism* 23(5), 660–682.
Lew, A.A. (2014) Scale change and resilience in community tourism planning. *Tourism Geographies* 16(1), 14–22.
Marjavaara, R., Müller, D.K. and Back, A. (2019) Från sommarnöjen till Airbnb: en översikt av svensk fritidshusforskning. In: Wall-Renius, S. and Heldt Cassel, S. (eds) *Turismen och resandets utmaningar*. Svenska sällskapet för antropologi och geografi, Stockholm, Sweden, pp. 53–77.
McIntyre, N. (2006) Introduction. In: McIntyre, N., Williams, D.R. and McHugh, K.E. (eds) *Multiple Dwelling and Tourism: Negotiating Place, Home and Identity*. CABI, Wallingford, UK, pp. 3–14.
McIntyre, N., Roggenbuck, J.W. and Williams, D.R. (2006) Home and away: revisiting 'escape' in the context of second homes. In: McIntyre, N., Williams, D.R. and McHugh, K.E. (eds) *Multiple Dwelling and Tourism: Negotiating Place, Home and Identity*. CABI, Wallingford, UK, pp. 114–128.
Mkono, M. and Markwell, K. (2014) The application of netnography in tourism studies. *Annals of Tourism Research* 48, 289–291.
Mody, M.A., Suess, C. and Lehto, X. (2017) The accommodation experiencescape: a comparative assessment of hotels and Airbnb. *International Journal of Contemporary Hospitality Management* 29(9), 2377–2404.
Müller, D.K. (2007) Second homes in the Nordic countries: between common heritage and exclusive commodity. *Scandinavian Journal of Hospitality and Tourism* 7(3), 193–201.
Müller, D.K. (2014) Progress in second-home tourism research. In: Lew, A.A., Hall, C.M. and Williams, A.M. (eds) *The Wiley Blackwell Companion to Tourism*. Wiley, Hoboken, New Jersey, pp. 389–400.
Nordin, U. and Marjavaara, R. (2012) The local non-locals: second home owners associational engagement in Sweden. *Tourism* 60(3), 293–305.
Norris, F.H., Stevens, S.P., Pfefferbaum, B., Whyche, K.F. and Pfefferbaum, R.L. (2008) Community resilience as a metaphor, theory, set of capacities, and strategy for disaster readiness. *American Journal of Community Psychology* 41(1–2), 127–150.
ONS (Office for National Statistics) (2020) Sub-National Tourism: A spatial classification of areas in England and Wales to show the importance of tourism, at county and unitary authority level, 2011 to 2013. Available at https://tinyurl.com/28hxp2uv (accessed 7 November 2020).
Panzer-Krause, S. (2020) The lost rural idyll? Tourists' attitudes towards sustainability and their influence on the production of rural space at a rural tourism hotspot in Northern Ireland. *Journal of Rural Studies* 80, 235–243.
Paris, C. (2009) Re-positioning second homes within housing studies: household investment, gentrification, multiple residence, mobility and hyper-consumption. *Housing, Theory and Society* 26(4), 292–310.
Pitkänen, K. (2008) Second-home landscape: the meaning(s) of landscape for second-home tourism in Finnish Lakeland. *Tourism Geographies* 10(2), 169–192.
Roberts, E., Anderson, B.A., Skerratt, S. and Farrington, J. (2017) A review of the rural-digital policy agenda from a community resilience perspective. *Journal of Rural Studies* 54, 372–385.
SCB and Tillväxtverket (2019) Gästnätter för samtliga hotell, stugbyar, vandrarhem, campingar, förmedlade privata stugor och lägenheter efter region och hemland. År 2008 – 2020. Available at: https://www.statistikdatabasen.scb.se/pxweb/sv/ssd/START__NV__NV1701__NV1701A/NV1701T910Ar/table/tableViewLayout1/ (accessed 4 February 2021).
Sheppard, V.A. and Williams, P.W. (2015) Systems-based and internal factors enhancing resort community resilience. Paper presented at the Tourism Travel and Research Association: Advancing Tourism Research Globally, 2015 TTRA International Conference, 15–17 June 2015, Portland, Oregon.
Smith, D.P. and Phillips, D.A. (2001) Socio-cultural representations of a greentrified Pennine rurality. *Journal of Rural Studies* 17, 457–469.
Suess, C., Woosnam, K., Mody, M., Dogru, T. and Sirakaya Turk, E. (2021) Understanding how residents' emotional solidarity with Airbnb visitors influences perceptions of their impact on a community: the moderating role of prior experience staying at an Airbnb. *Journal of Travel Research* 60(5), 1039–1060.

Tuulentie, S. and Kietäväinen, A. (2020) New rural community? Narratives from second home owners about everyday life in a tourist region in Finnish Lapland. *Sociologia Ruralis* 60(2), 357–374.
Vepsäläinen, M. and Pitkänen, K. (2010) Second home countryside. Representations of the rural in Finnish popular discourses. *Journal of Rural Studies* 26, 194–204.
Walters, T. (2016) Using thematic analysis in tourism research. *Tourism Analysis* 21(1), 107–116.
Williams, D.R. and Van Patten, S.R. (2006) Home and away? Creating identities and sustaining places in a multi-centred world. In: McIntyre, N., Williams, D.R. and McHugh, K.E. (eds) *Multiple Dwelling and Tourism: Negotiating Place, Home and Identity*. CABI, Wallingford, UK, pp. 32–50.
Wilson, G. (2010) Multifunctional 'quality' and rural community resilience. *Transactions of the Institute of British Geographers* 35(3), 364–381.
Woods, M. (2010) Performing rurality and practising rural geography. *Progress in Human Geography* 34(6), 835–846.

8 Local Commitment and Withdrawal in the Wake of Conspicuous Airbnb Place Dynamics on a Cold-Water Island

Michael Röslmaier[1]* and David Albarrán[2]
[1]*Mid-Sweden University, Östersund, Sweden;* [2]*Complutense University, Madrid, Spain*

Abstract

In recent years, Airbnb has gained tremendous attention in sense-of-place and community-resilience research. Yet, we find few scholarly efforts to relate particular issues of community resilience to a sense of place that emerges in the wake of conspicuous place dynamics of Airbnb beyond the urban context. In this case study on Heimaey, Iceland, we use geographic information systems (GIS), AirDNA data on Airbnb, and in-depth interviews to address this research gap. We find that as professional Airbnbs concentrate in the 'downtown' (town centre) area, residents mourn over lost pasts with a heightened sense of place that fuels intentions to withdraw. The isolated, casual, Airbnb landscape in the 'residential west', meanwhile, still allows locals to retain stewardship over place meanings, identities and community structure, which translates into increasing commitment out of fear of losing these privileges in the near future. This chapter encourages future research to become concerned with Airbnb developments in non-urban environments where sense of place and community resilience are both intrinsically linked and negotiated by different types of host activities in different locations.

Keywords: Airbnb, tourism, community resilience, place identity, sense of place, Iceland, GIS

Highlights

- This research investigates the spatiality and character of distinctive Airbnb developments in a non-urban context and relates such developments to issues of sense of place and community resilience.
- We show that concentrations of professional Airbnb offerings leave behind emotionally disconnecting traces in the physical environment to which residents respond with strong intentions to withdraw from place and community.
- This stands in sharp contrast to isolated, casual and semi-professional Airbnb activity that contributes to a heightened sense of place that fuels local commitment to both place and community.
- Sense of place can serve and hinder the goal of achieving community resilience in a non-urban environment that finds itself amid conspicuous Airbnb place dynamics.

*Michael.Roslmaier@miun.se

 Peer-to-peer Accommodation and Community Resilience (eds Anna Farmaki, Dimitri Ioannides and Stella Kladou)

DOI: 10.1079/9781789246605.0008

Introduction

Airbnb is a home rental platform for home-owners to make their properties available to the global visitor market, renting out *authentic* local homes (Airbnb, 2020a) meant to portray the uniqueness of a community in the destination. Today, we see a disproportionate growth of **superhosts**, who, according to Airbnb, are shining examples of highly experienced and dedicated, albeit often absent, hosts who provide extraordinary experiences and outstanding hospitality by means of multiple listings that serve as 'tropes of home' (Roelofsen and Minca, 2018). In many cities, this professionalization and economization of housing effectively transforms places, historical buildings and entire communities of homes into reflections of outside interests by turning small-scale personal properties into tourism capital (Booth, 2015).

The relationship between Airbnb and place transformation has been widely discussed in the urban literature (see Guttentag, 2019). However, little is known about the implications that short-term hosting activities have for remote, rural and small-scale communities, which are in high demand, yet have little to no background in tourism (UNWTO, 2016; UNWTO *et al.*, 2018). This, for instance, applies to many fishing communities in Iceland. Tourism may have brought 'resilient Iceland back from the brink' (Partington, 2018) following the country's downfall during the economic crisis in 2008. Airbnb, however, is increasingly recognized as a force that pushes the sociocultural carrying capacities to their tipping points since short-term rentals replace local residents with global hosts and guests (Jóhannesson, 2015; Kristjánsdóttir, 2016; *The Telegraph*, 2019). Thus, while tourism can serve as a vehicle of opportunity for such communities to escape isolation, achieve economic diversification and join the global economy (Bertoncello, 2015; Albarrán Periáñez, 2020), the manner in which Airbnb operates can act as a barrier to reach the goal of achieving social sustainability and community resilience (Mermet, 2019).

It is, thus, urgent to look into the implications of Airbnb developments on places, society and everyday life (Dredge and Gyimóthy, 2015). While statistics tell us only how many visitors arrive in large, historically significant tourist-friendly spaces each year, they tell us little about the wear and tear on everyday streets and local life. In this study on the small fishing-island community of Heimaey, Iceland, we utilize the concepts of place identity, sense of place and place attachment, in order to unravel and relate local reactions to different Airbnb developments in certain places of meaning, to issues of community resilience. To date, little is known about the wear and tear that different volumes and types of Airbnb activities, such as high concentrations of professional home hostings, leave behind in the physical and social environment, particularly in rural and peripheral regions (Mair and Reischauer, 2017). In fact, Norris *et al.* (2008) argues that changes to the physical environment bear implications for a sense of place and community, as well as place attachment, all of which underlie 'citizens' efforts to revitalize a community' (p. 139). While currently understudied, Airbnb developments, place transformations and community resilience may go hand in hand. Thus, we address these research gaps by drawing on the connotations of Airbnb as a conduit for social change through place transformations. In doing so, we avoid absolutist rhetorical and normative statements that reduce Airbnb to *one* entity of urban and economic concern for gentrification, sense of place and place attachment as always positive for sustainability, while community resilience functions as a mere collective representation of a desired *state of the past* (see Masterson *et al.*, 2017). Although this research was developed prior to COVID-19, it is still relevant to identify the vulnerabilities of small-scale communities to transformations of places in the context of Airbnb development.

To this end, we map geographical and technical information on Airbnb activities through geographic information systems (GIS). In our field observations, we collected photographs and 37 in-depth interviews in order to link the spatialities and character of Airbnb activities to changing components of places (e.g. place identity) and reveal their implications for community resilience.

Theory

Airbnb, place transformation and community resilience ...

Many scholars recognize Airbnb as an agent capable of accelerating the touristification of places and communities, eventually becoming a matter of social resilience (Dolnicar and Zare, 2020; Garay *et al.*, 2020; Stors, 2020). In a Berlin neighbourhood, for instance, Stors (2020) argues that hosts give rise to new tourism areas by negotiating the discursive (e.g. implying new meaning) and material (e.g. appearances of short-term rentals) aspects of places. In some ways, the expected benefits of an Airbnb-tourism economy are promising; it features cultural exchange and the potential to create new communities, local spaces and a hyper-local economy. Simultaneously, ideas of belonging and stranger status may become complicated amid a disproportionate presence of corporate-driven home rentals vis-à-vis home-sharers and long-term residents in rural areas. Tourism, thus, can affect the ways in which locals relate to their environment by commodifying notions of scenery and aesthetic readings of land according to the ideas of hosts and expectations from guests (Hernández Ramírez, 2008; Albarrán Periáñez, 2021). That, in turn, can create conflict between different interest groups, such as Airbnb investors and residents living close by, with repercussions for social (community) resilience.

Community resilience emerged as an important framework to assess a group's willingness and ability to cope with external interferences capable of threatening community functions, identities and places by disrupting the 'critical triangle of capitals' (Zwiers *et al.*, 2018, p. 285), that is the social, economic and physical (Wilson, 2012). Thus, resources, actions and social values are centre stage in resilience literature (Buikstra *et al.*, 2010). Yet, issues of identity and notions of place remain surprisingly neglected and undertheorized, in particular when it comes to Airbnb developments in non-urban areas. It is thus important to identify the relationship between Airbnb and changes to place components and the manners in which they, eventually, inform human action, social (ex-)change, as well as meanings, identities and histories that are attached to such places (Folke *et al.*, 2010). Eventually, these components come together in the concept of sense of place.

... from a sense-of-place perspective ...

In this chapter, we define sense of place as an intangible resource constructed by different spheres of cognition and emotion that is inherited by people's meanings and attachments of a particular setting (Jorgensen and Stedman, 2001). With place attachment we refer to people's 'emotional bond to a particular geographic space' (Altman and Low, 1992; Zwiers *et al.*, 2018, p. 283). The mosaic of current Airbnb rentals, meanwhile, suggests that we have to acknowledge an instrumental connection between people and places (Stokols and Shumaker, 1981; Roelofsen and Minca, 2018) that can be captured through interview items such as 'This is the best place to do the things I enjoy' (Jorgensen and Stedman, 2001). This relates back to Proshansky's (1978) conceptualization of place identity as 'those dimensions of self that define the individual's personal identity in relation to the physical environment' (p. 155). We form and perform *the self* through engaging with symbolic questions like 'Who am I?' (Sack, 1997) and 'Does "this place" really reflect the kind of person I am?' (Jorgensen and Stedman, 2001). Thus, much like the concept of *place* itself, our senses and attachments to place identities and meanings are as much personal as they are social, environmental, material and immaterial.

In Airbnb research we can, thus, expect that uncontrollable changes to our personal relationship with places have been translated into different intentions and capacities to adapt as a collective. While some may withdraw from places when certain Airbnb developments make spaces impersonal and unaffordable, others may decide to 'stay and protect what we cherish most in our communities and to invest time, energy, and money to improve that with which we are dissatisfied' (Perkins *et al.*, 2002, p. 41). This suggests that when in flux, our emotional bonds to meaningful places inform changes to the relationship between the 'I' and 'We', which then affects personal and social capacities and intentions for maintaining and developing community resilience (Brown and Vergragt, 2016).

Often, scholars in sense-of-place and place-attachment research fall into the trap of assuming that all people with strong personal attachments to places in flux will naturally mobilize to protect their place from certain threats (Stedman, 2008). It is here where place meaning becomes pivotal to this study. Place meaning refers to the: (i) descriptive, sociopsychological cognitions about what a place is (e.g. empty, sad, warm); (ii) a symbolic and interpretive stance of what a place meant and means (e.g. reminders of history, personal and collective achievement); and (iii) on the character of a place (i.e. tourist place, local place) (Manzo, 2005; Brehm *et al.*, 2013; Lyon, 2014). As place meanings are also central in the world of Airbnb, levels of personal and shared interests relating to place, history and roots are being negotiated, which can either strengthen or weaken our attachments to places as well as community ties and commitments to community resilience (Mermet, 2019). In other words, by following Berno and Ward (2005), we are looking at Airbnb-related social change through the transformative dimensions of the symbolic (place identity, sense of place), functional adequacy, and the reinvention of places of meaning through physical adaptation (e.g. change from residential home to tourist rental).

... in the context of rural, peripheral, small-scale communities

This becomes especially important when investigating remote, isolated, rural and small-scale fishing communities like Heimaey – one of Iceland's Westman Islands – where we can expect strong internal ties to place and between community members owing to shared histories, traditions, memories, attachments and tendencies of long-term, cross-generational occupancy (Altman and Low, 1992; Walker and Salt, 2006). These factors are not only incremental to the development of community and place pride, but also eventually fuelling community participation. They are also subjected to current Airbnb dynamics as soon as home hostings invite nuanced land-use practices, new local behaviours, and wear and tear to streets, buildings, sites and places (Altman and Low, 1992; Muller *et al.*, 2018; Stors, 2020). It is thus important that exogenous interventions, such as Airbnb hosting activities, serve residents as a resource rather than as barriers: (i) to retain the 'same function, structure, identity, and feedback' (Walker *et al.*, 2004, p. 5); and (ii) to relationships with community and places (see Füller and Michel, 2014; Gutiérrez *et al.*, 2017).

Ultimately, we utilize the concept of sense of place in order to draw out connections between certain Airbnb developments, place-specific wear and tear, and issues relevant to the goal of achieving community resilience in the case of Heimaey, a tradition-oriented, isolated, small-scale island community.

The Case of the Westman Islands

A mere 20-minute flight away from Reykjavik (see Fig. 8.1) lies the volcanic archipelago of Vestmannaeyjar (Westman Islands). Among the 15 islands, Heimaey stands out as the largest and only inhabited place. With an estimated population of 4500, EyjaTours (2019) describes Heimaey as 'Iceland's best kept secret, with a rich history, unique landscape, and heart-warming, welcoming inhabitants'. However, there is also a darker side to the island's past. This largely has to do with the island's vulnerability to volcanic eruptions and its physical isolation. In 1973, large parts of the island had to be evacuated following the eruption of its largest active volcano (VWI, 2020). Five years later, most residents decided to move back and rebuild places and their community from scratch despite knowing that nothing would ever be the same again. As large parts of the island are covered by volcanic ash, opportunities for housing developments remain severely limited. Moreover, residents have for long endured relative isolation from the mainland. Many could not afford flights. Instead, they relied on favourable weather conditions that allowed ferries to commute between the island and the mainland once per day.

Thus, natural forces and the geographic location and characteristics of the island have long determined where, how and when people could work and settle. So far, from an economic point of view, their high share of the nation's fishing quota also rendered them relatively stable and independent in times of crisis (e.g. the economic crisis of 2008). Although fishing remains the biggest source of income in Iceland,

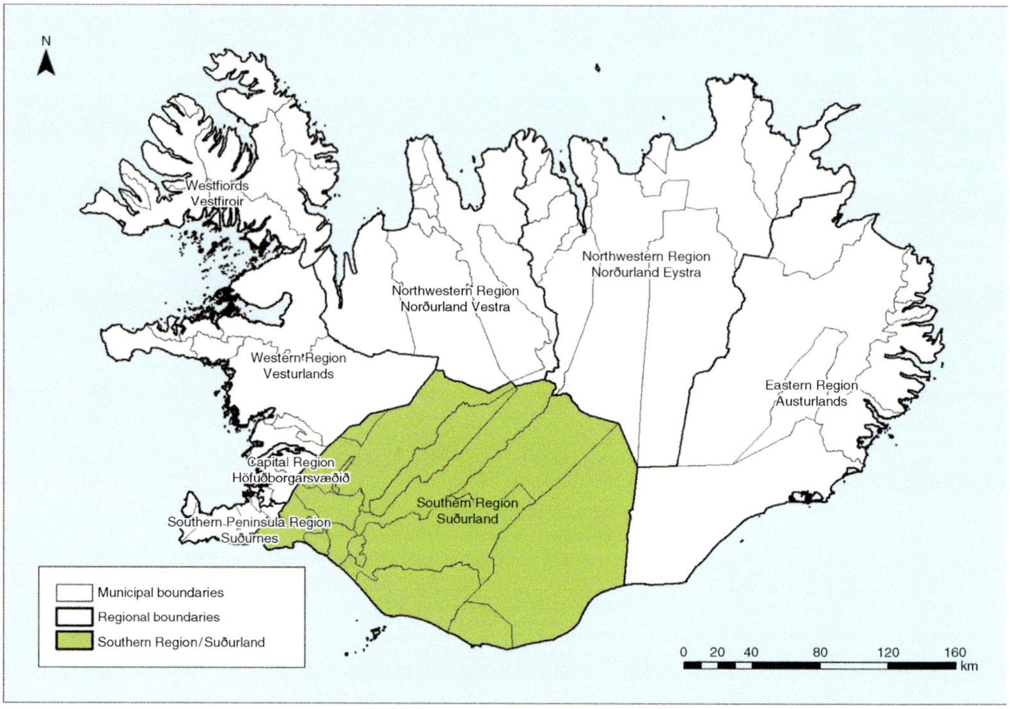

Fig. 8.1. Location of Vestmannaeyjar (Westman Islands), of which Heimaey is the largest and only inhabited island, south of the Southern Region, Iceland. Adapted from cartographic data from Landmælingar Íslands, 2019.

and particularly Heimaey, over the past 10 years tourism has increasingly developed into a serious alternative for the islanders (Óladóttir, 2018). Since then, the island's transformation into a visitor destination has left marks in the composition and characteristics of communities and their places. For instance, this transition motivated the emergence of enhanced services and infrastructure capable of catering to the image and appeal of the destination in a global tourism market. Tourists and locals alike can now benefit from a more pleasant, frequent ferry service.

Major attractions have been gradually appearing, such as an impressive aquarium built with international capital, a local brewery, restaurants of high repute and tour operators. At the same time, cultural and social practices are closely tied to the histories and identities of places on the island and the annual use cycle of its natural resources. Together, this has boosted Heimaey's global reputation, resulting in rising numbers of cruise ships that now count the island as one of their ports of call and a boost to its attractiveness to alternative global tourists aiming for homestays.

Visitors today can choose from an array of Airbnb properties that have significantly increased the frequency of overnight stays, complementing and competing with the one hostel and hotel that operates on the island (Airbnb, 2020c). Ultimately, Heimaey is no longer merely a lonely fishing outpost and shelter for local residents. Through transnational capital, local entrepreneurship and increased numbers of international visitors, the island is emerging as an attractive tourism destination with services such as Airbnb that have introduced a variety of new land-use practices, local behaviours, and wear and tear previously unknown to its community.

Methods

In this study we follow a typical case-study design and mixed-method approach that helps us to gain an in-depth understanding of a particular geographical area and community in the context of Airbnb development (Yacuzzi, 2005; Álvarez and San Fabián, 2012). First, we gathered and mapped data on current Airbnb activities with the help of ArcGIS 10.4.1. Due to the proprietary nature of Airbnb data, we purchased this information from AirDNA (data on the Westman Islands is not publicly available), an intermediary service that on a daily basis scrapes information from Airbnb's listings. For the purposes of our study, AirDNA used algorithms to estimate important data on 104 active listings that were active at the time of our study, including: (i) location; (ii) availability; (iii) average price and revenue; (iv) ownership; and (v) size. These categories allowed us to differentiate and typologize listings in accordance with the Icelandic Home Hosting Act. This act allows owners to list any residence as long as they do not exceed five rooms/separate spaces or beds for more than ten guests over a combined 90 days and a consecutive 30-day period per listing, and in total do not generate more than ISK (Icelandic króna) 2,000,000 (approx. US$15,000)/year as a host. In cases where Airbnb listings do not satisfy all these criteria, hosts must apply for a licence, pay taxes and follow, among others, certain rules of hygiene, safety, check-in routines, booking procedures and accessibility (Airbnb, 2020b). Certain requirements necessitate changes to physical appearances of buildings, such as emergency ladders, key boxes and welcome signs. We can thus expect visible contrasts between the effect of concentrations of professional touristified homes and casual home-sharings on the physical and social environment of places.

For this, we conducted on-site observations over 3 weeks during the summer of 2019. As we traversed the town on our own but also with locals, we identified signs of visible changes to the appearance of buildings (e.g. key boxes, multiple escape ladders, decorations, building style) that we captured in photographs and field notes. According to several scholars, photographs can highlight how different tourism-related developments like Airbnb are visibly embedded in the local landscape (Gual *et al.*, 1986; Russo and Richards, 2016). Moreover, we identified two contrasting types of hosts (i.e. the professional and the casual home-sharer). One of us stayed, over a period of 1 week at a time, with each of these respective hosts. This allowed us to gain an in-depth understanding of the locals' relationship with places experiencing change amid different Airbnb developments.

By means of snowball sampling, we identified 35 additional participants for the purpose of conducting face-to-face, in-depth, semi-structured and open-ended interviews. Each participant had to have lived on the island prior to the proliferation of Airbnb in 2013 in order to derive and emphasize voices of reactions to change related to this phenomenon. Among our interviewees were Airbnb hosts, business owners, politicians, public-sector employees, experts (an historian, a tourism management representative and a real estate agent) and permanent residents (see Table 8.1). Due to the sensitivity of this topic, we refer to all interviewees as residents with pseudonyms in order to ensure anonymity. We recorded, transcribed and thematically analysed each interviewee with the help of Dedoose, a qualitative data analysis software. According to Benediktsson *et al.* (2011), a thematic analysis requires a systematic examination and interpretation of data in accordance with a theoretical frame.

Ultimately, these techniques helped us to identify and locate concordances between Airbnb developments, issues of (sense of) places in change, and their consequences for community resilience.

Quantitative Results: Mapping Airbnb on Heimaey

The data we obtained from AirDNA shows an immense increase of Airbnb listings over the 4-year period 2014–2018, up from five listings in 2014 to over 100 by 2018 (see Table 8.2). Currently, only two out of 104 listings fall under the category of **shared rooms** within private homes. This is surprising given that it is safe to assume that this category manages to convey a more *authentic* sense of *this* place as opposed to uninhabited entire-apartment/house rentals run by either companies or individuals living elsewhere on the island or abroad. Moreover, the vast majority of listings are located in and around the old town centre (see Fig. 8.2). Meanwhile, rentals of entire houses appear to be increasingly popular in residential areas closest to downtown (the town centre).

Table 8.1. List of interviewees.

Pseudonym	Role[a]	Pseudonym	Role[a]
Anja	Permanent resident	Luna	Business
Ben	Business	Manuel	Business
Chloe	Permanent resident	Maria	Host
Christian	Expert	Martha	Expert
Claudia	Business	Miriam	Permanent resident
Daisy	Public-sector employee	Olga	Permanent resident
Daniel	Expert	Paul	Permanent resident
Dieter	Host	Quinn	Business
Emil	Politician	Richard	Public-sector employee
Felix	Public-sector employee	Roberta	Permanent resident
Hans	Host	Siggy	Business
Ingo	Business	Thomas	Permanent resident
Jenny	Business	Tom	Politician
Jessy	Permanent resident	Udo	Business
Klaus	Permanent resident	Vanessa	Permanent resident
Kristin	Politician	Walter	Permanent resident
Lena	Permanent resident	Willy	Public-sector employee
Lennard	Business	Yvonne	Business
Lisa	Host		

[a]Experts include an historian, a tourism management representative and a real estate agent.

Table 8.2. Number of Airbnb listings per year on Heimaey by type. Based on data purchased from AirDNA, 2019.

	New listings				
Year	Entire houses/ apartments	Private rooms	Shared rooms	Total new listings	Total supplies[a]
2013	2	3	0	5	5
2014	0	0	0	0	5
2015	5	9	0	14	19
2016	8	12	0	20	39
2017	13	20	0	33	72
2018	20	10	2	32	104
Total	48	54	2	104	104

[a]Total supplies is cumulative, including pre-existing listings as well as all the new listings for each year.

Heimaey's new waterfront, which is characterized by tourism-related business and housing developments that – according to the experts we spoke to – are unaffordable for most persons, also contains a substantial Airbnb presence. According to our interviewees, the accumulation of Airbnbs in the downtown area corresponds to the growing demand by residents for novelty and sophistication. In other words, when these residents leave the downtown area behind in order to move to the surrounding neighbourhoods or the new, modern and luxurious apartments at the waterfront they leave behind properties that can be converted into Airbnbs. This further allows and motivates the upsurge of professional Airbnbs, as real estate investors find it easier to stem the immense costs of restoring properties and upkeeping them in a very attractive location for visitors who, eventually, pay their bills.

Thus, it is no surprise that the average annual income derived from each listing is as high as US$11,303. A total of 36 rentals, all of which are entire homes and apartments, yield more than US$15,000 (see Table 8.3), with the top seven

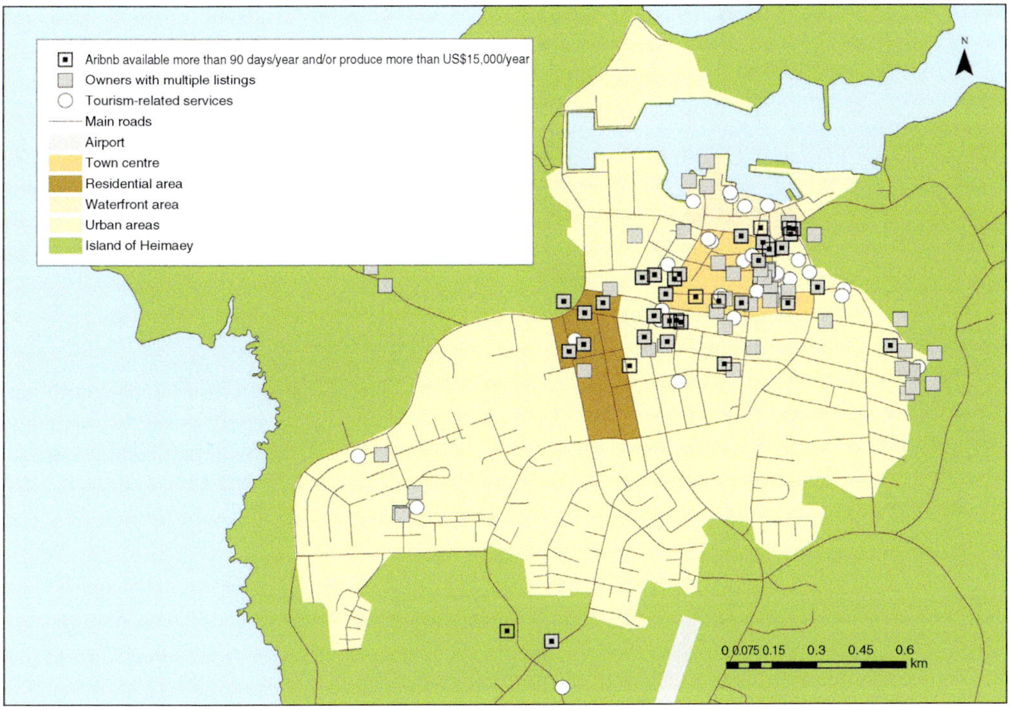

Fig. 8.2. Location of concentrations of professional Airbnb tourism activity in Vestmannaeyjabær, the only town on Heimaey. From observations, experts and AirDNA.

Table 8.3. Basic quantitative information of the Airbnb offerings in Heimaey in 2019. Based on data purchased from AirDNA, 2019.

Feature of the Airbnb offering	Number of properties
Type of accommodation	
Entire houses/apartments	48
Private rooms	54
Shared rooms	2
Capacity	
Less than or equal to four guests	70
Between four and ten guests	34
More than ten guests	0
Annual revenue	
Less than or equal to US$15,000	68
More than US$15,000	36
Availability (days/year)	
Less than or equal to 90 days	37
More than 90 days	67

totalling over US$30,000. A total of six hosts operate approximately 50% (N = 50) of the short-term rentals; each owns at least four and up to 21 houses or apartments. The annual mean income per host, meanwhile, totals US$26,712.34. This high number, however, is driven up by 16 hosts who earn up to US$280,000 and together average US$58,846.06. In comparison, the mean income of hosts owning only one property is as low as US$8062.27. This underlines the profitability of running homes as tourist-stays on a professional, profit-oriented basis.

Like in urban areas, the most profitable offerings on the island are in the heart of the only town on the island, Vestmannaeyjabær, and at the new waterfront (see Fig. 8.2). Although none of them supply more than five rooms and/or beds for more than ten guests, 67 properties are available for more than 90 days a year (134 days on average), the vast majority operating downtown. This suggests that 37 out of 104 properties fall within the Icelandic Home Hosting Act and are considered to be **residential homes**. Furthermore, similar to results found in larger cities, the majority of professional, short-term rentals are located in close proximity to other tourism-related establishments, primarily in downtown and along primary shopping streets (see Ioannides *et al.*, 2019). In the residential neighbourhood of West Heimaey we can meanwhile observe a much more isolated, casual and semi-professional Airbnb landscape (see Fig. 8.2). Eventually, this contrasting landscape of Airbnb (i.e. in terms of numbers, type of ownership and rental) likely leaves behind different traces in, and implications for, the physical and social environment in these different locations.

Qualitative Results

Downtown Heimaey: voices of withdrawal amid professional Airbnb concentrations

Almost instantly, we observed evidence of Airbnbs around the downtown area. The main streets looked forsaken: many buildings were in bad shape, and business fronts had clearly been abandoned amid renovated, sterile-looking houses of which the majority can be easily identified as tourist accommodation. A few streets over, we saw flourishing, renovated houses – colourful, lively and with a decorative charm – that are currently inhabited by local residents. 'We are all proud of these houses ... unlike those [Airbnbs].' (Maria). She pointed at a building unlike any other, explaining that despite being renovated, this is nothing more than 'a business that decided to turn this home into an attraction for tourists' (Maria) (Fig. 8.3). Unsurprisingly, nearly all interviewees lamented this type of Airbnb activity, referencing the deterioration of local charm, historical buildings, local lifestyle, and places of personal and social significance that got lost in translation in the modernization process. According to locals, each building used to tell a part of 'our story ... and everything we have been going through together' (Thomas). Olga believes that buildings are meant to serve as an illustration of how 'deeply we are connected to this place and to our history'. Yet, their meanings are now becoming lost in translation as the histories of places of significance are not only scrambled into Airbnb charms but also serving the needs of Airbnb hosts and guests rather than those of long-term residents and locals. In that respect, Airbnb, eventually, introduces urban traits to rural places (see Domínguez and Scarnato, 2017; Gutiérrez *et al.*, 2017). This led our study participants to respond with anger, disappointment and disdain.

Every sign of Airbnb activity, such as key boxes, escape ladders, guests welcome signs and numbered doors, was described as 'sterile', 'unfamiliar', 'disgusting', 'estranging' and 'sad'. Although most interviewees were happy to know that someone is invested in renovating the

Fig. 8.3. Airbnb contrasts in downtown Heimaey. (Note the modernized, newly renovated properties for tourists, identifiable as sterile-looking buildings with numbered doors.) Photo courtesy of the authors.

many old, run-down buildings in this area, the fact that Airbnb was the impetus for renovation stifled their happiness. 'We all know that most of these houses become Airbnbs ... nobody lives there ... and nobody is going to' (Paul). In fact, downtown is also the 'perfect location for Airbnb. You need less than a minute to get anywhere you want to go as a tourist' (Lena). Just like in large cities (see Ioannides *et al.*, 2019), short-term rental location in proximity to tourism-related infrastructure also matters in a destination as small as Heimaey. In the end, however, 'this whole Airbnb-thing has changed our city centre' (Maria).

Downtown Heimaey first emerged as the centre of life back in the days when locals repopulated the island following the volcanic eruption (Martha). Everything, places, its buildings and their flair, no matter how *rundown* they may appear today, still tells a story of personal and collective achievement in overcoming one of the 'biggest disasters in our history' (Kristin). 'It took a lot of effort to make this island likeable again and to save many histories, meanings, and identities ... this is what made and kept us strong as a community' (Martha). After the eruption in 1973, it was here where locals found each other again, established social spaces for reconnecting, and reinvented themselves (Manuel). Today, places downtown are primarily shifting towards the new waterfront. According to most interviewees, the old fishing market is what 'everyone here misses the most ... it was the only fishing market on an island that is all about fishing ... can you believe that? ... here, we felt at home' (Martha). According to Martha and Maria, it is sad to see traditional shops in the downtown area gradually disappearing. That means places of meaning continue to become increasingly rare as many businesses and houses around the downtown area are designed to serve the needs of the tourist market over those of locals (Klaus). 'We feel left out' (Miriam). We feel that our histories become lost in translation, and 'what is a place without people living in it... without a community that makes sense out of a place?' (Martha). In the end, it became evident that local residents connect current Airbnb developments to place compositions and community changing in a personal, social, material and immaterial sense (Raymond *et al.*, 2010).

In other words, professional Airbnb hosts negotiate and, to a large extent, disconnect locals' sense of place and place identity, severing connections between people and this particular geographical area where they once shared strong socio-emotional bonds. The high concentrations of such contrasting home-hosting styles further appear to complicate feelings of belonging while also increasing stranger status. According to Martha, there is:

> little we can do about it ... that all these houses are bought up by investors nobody knows ... people who don't know what this place means to us ... all we can do is watch and do the best we can to deal with it ... or not ... I think that this is why many decide to leave.
>
> (Martha)

Sooner rather than later, 'everything and everyone we care about ... disappears' (Anja; Lennard). Streets turn 'emptier, darker, and emotionless' (Hans) as people with whom locals shared stories and histories move away from the island (Anja; Ingo). The fact that, today, downtown 'tells the story of a rich stranger running a "hotel"' (Manuel) makes us 'want [to get] the hell out of here' (Anja; Miriam). These are references to increasing feelings of withdrawal that connect Airbnb to matters of sociopsychological cognitions, symbolic values and the changing sense of character of downtown (Manzo, 2005; Brehm *et al.*, 2013; Lyon, 2014). Airbnb drains emotional capacities, eventually resulting in personal intentions to detach from places and a community *of increasing strangeness*. In other words, the everyday wear and tear that this concentration of professional Airbnb leaves behind in the built/physical environment motivates intentions of withdrawal, eventually informing social change. According to many scholars, this is unlikely to serve the goal of achieving and building meaningful community resilience, in particular in the context of non-urban environs (Sack, 1997; Jorgensen and Stedman, 2001; Masterson *et al.*, 2017; Zwiers *et al.*, 2018).

Voices of commitment: the case of casual home hosting in West Heimaey

Reflections about developments in West Heimaey were markedly different. In this residential area, our participants appreciated a more isolated, casual and semi-professional home-hosting landscape. Moreover, they explicitly recognized it as less destructive to place meanings, identities and their community as opposed to the professional concentration downtown. Hans believes that this is because most Airbnbs in this area are run by respected community members with strong ties to both place and communities. Hans intends to maintain this situation to ensure the endurance of strong emotional connections between community members and their relation to place.

> I know people here appreciate that I try everything to change as little on the outside of this house as possible. I can earn money by renting out an authentic home where everyone profits. My guests welcome this ... and so [do] our neighbours ... because they still feel like home.
>
> (Hans)

Moreover, this situation also makes us feel like we still 'have a say about what is happening with this place and us ... it is a real community because everyone is on the same page ...; to prevent what is happening to downtown [from spreading] here' (Richard). One time, Kristin highlights, 'someone wanted to build many cottages here. People became very upset about their architecture ... they voted against it'. Eventually, the investor decided to respect their wishes to 'make them appear more local' (Kristin). In the end, she argues, we all know that 'when you change too much too fast, this place becomes nothing more than a summer resort too ... we definitely don't want that to happen'. 'We don't want to feel like strangers in our own city ... and we don't want to make others feel that way either' (Roberta). These quotes suggest that on Heimaey, semi-professional, casual and isolated Airbnb activities are less destructive to the essential, instrumental connections that community members share with each other and places, eventually translating into intentions of commitment rather than withdrawal (Stokols and Shumaker, 1981; Masterson *et al.*, 2017).

The fear of seeing high concentrations of professional Airbnbs and the subsequent related changes to the environment in downtown Heimaey raises individual and collective awareness of 'who I am', 'who we are', 'what is going on', 'what a place means', and 'what community is or should be about' (see Sack, 1997; Jorgensen and Stedman, 2001). The local voices we hear on Airbnb developments in West Heimaey, dissimilarly, embodied commitment rather than withdrawal: residents prioritized efforts and intentions to stay and protect everything they cherished about their communal places and community in general. This collective effort is what it takes to maintain community resilience (Perkins *et al.*, 2002). Moreover, in order to maintain community resilience, it takes collective efforts 'to convince investors that ... at some point, enough is enough' (Tom). Depending upon the surrounding context, Airbnb can either weaken or strengthen community resilience. In this case, Airbnb 'makes us grow closer together ... Stronger ... and more aware of what this place and everyone here actually means to us' (Olga). 'I don't think this would be the case when things here would get out of [our] hands' (Hans).

Moreover, our interviews show that Airbnb can contribute to building and strengthening locals' sense of place and place attachment that, eventually, promotes community resilience. In fact, different community stakeholders (e.g. future investors, current hosts, permanent residents and politicians) appear to engage in negotiations regarding, and relating to, issues of place identity, place meaning and community. Yet, there is also evidence that suggests growing intra- and intergroup competition and interest conflicts based on different ideas of 'what is supposed to become of this place' (Thomas), which can both strengthen or weaken personal and social capacities and intentions to cope with (fear of) Airbnb developments and downtown-like change (see Brehm *et al.*, 2004).

Withdrawal and commitment: a matter of capacity and stewardship

According to our interviews, Airbnb undoubtedly bolsters the tourism industry and certain dimensions of community well-being, in that it brings 'the beauty back to this place' (Maria). 'Today, there are so many new things to do in the summer' (Claudia), 'new beautiful housing developments' (Thomas), 'different work environments' (Richard), and 'so many nice people [visitors] ... all this makes us feel less isolated ... like living in a big city' (Martha). Home hosting also helps many 'to secure down payments of mortgages' (Hans; Jenny). All in all, locals would 'expect more young people to move here to realize all these great opportunities' (Richard) or 'to see [their] population finally growing again' (Martha).

Contrastingly, 'housing is just not affordable anymore as it was before' (Richard). 'We can only build so many houses here' (Ben), many of which are now owned by Airbnb investors 'who can more likely afford them' (Maria). Yet many of those investors are not living on the island, and 'many feel that the community shrinks faster today than it ever did before ... and I don't think that will change soon' (Martha). Clearly, it is hard to 'make it right for everyone. If this place changes too much like downtown, people become upset and leave. If you don't change too much, tourists become upset ... if we upset tourism, we upset everyone' (Hans).

Thus, Hans believes professional home-sharing activities are counterproductive to the goal of maintaining and developing strong emotional ties to place and community members. This sense of connection is a crucial foundation for achieving community resilience (Altman and Low, 1992). Instead, things have become touristified and characterized by strangers and strangeness (Jessy; Chloe; Vanessa; Udo; Siggy; Yvonne) amid growing housing issues and gentrification. 'If we cannot get this Airbnb thing under control, this may well be the beginning of the end of our community as we know it ... just look at what is happening elsewhere in this country' (Daisy).

> You know ... many of these hosts live in Reykjavik but own many apartments and houses here so they can become rich from renting [them] to tourists. This is not fair. They are not part of this community. They give us little back, except for more tourists, of course ... but they need to understand that life is not all about money or tourism ... even if we need these things.
>
> (Kristin)

It feels like:

> they decide what happens to us ... what we are ... but we have to decide over our own fate and history ... [our] own community ... With everything that is going on here in downtown, I don't see this happening ... especially if things get worse [or spread].
>
> (Quinn)

In that sense, community resilience indeed comes down to retaining stewardship over place meanings, community structure and identity (Walker *et al.*, 2004). Most participants agree that this is easier to achieve through isolated, casual, semi-professional home hostings that are less intrusive to the physical and social environment. Hence, we need to motivate and support more locals to occasionally share their homes during the season rather than investments by strangers who seek to acquire multiple houses and apartments in such a restricted location (Thomas). 'If we can get to this point, then we can expect to see more meaningful efforts from people to move here, stay here, and to involve [themselves] in this place and community again' (Kristin).

Discussion and Conclusions

This study shows that Airbnb activity can accelerate social change and the touristification of destinations far off the beaten tourist track. Despite Heimaey's small size, we found high concentrations of professional Airbnb activity in the downtown area analogous to concentrations found in bigger cities (Guttentag, 2019; Ioannides *et al.*, 2019). This stands in sharp contrast to the residential area in West Heimaey, where home hostings are isolated, casual and semi-professional. According to our participants, these two home-hosting styles have had distinctive impacts on Heimaey's physical environment, with differing repercussions for the social environment and implications for community resilience.

When discussing Airbnb developments on Heimaey, participants reacted to the professional concentration of short-term rentals in downtown with strong antipathy, as they hold such Airbnb rentals somewhat responsible for the irreversible alienation and touristification of downtown. As a matter of course, residents express intentions to withdraw and disconnect from both place and community. These reactions stand in sharp contrast to the more isolated, casual, Airbnb landscape of West Heimaey. In fact, it is the fear of professional home-sharing expansion into this neighbourhood that translates into a heightened sense of place and, eventually, into intentions to redevelop stronger personal and social commitments and attachments to both place and community.

It is safe to assume that, unlike commitment, withdrawal is counterproductive to the goal of achieving, maintaining and developing collective capacities to cope with future place-related stressors from professional Airbnb expansion (Sack, 1997; Jorgensen and Stedman, 2001; Masterson *et al.*, 2017; Zwiers *et al.*, 2018). Rather, by leaving Heimaey and its community behind, people make more spaces available for investors to purchase and transform into tourist rentals and attractions. Although many appear to appreciate Airbnb's role in 'reviving' the town and improving the appeal of some houses, they do not approve of losing stewardship over community structures, place meanings and identities in this process. Here, we find that Airbnb serves as a platform that empowers hosts to negotiate community resilience by encouraging (West Heimaey) and imposing (downtown) new local behaviours, land-use practices, place meanings and identities, as well as community structures (Stors, 2020). Our case study shows that this, eventually, can either make (i.e. commit) or break (i.e. withdraw) community resilience (Zwies *et al.*, 2018).

Thus, Airbnb development, sense of place and community resilience in Heimaey go hand in hand and should, as such, deserve recognition in governmental decision making. For many years, according to a local politician, their focus was on achieving tourism growth by improving the aesthetic appeal of the point of entry (i.e. the waterfront) for both potential visitors and residents, despite knowing about the people–place dynamics that unfold amid Airbnb expansion elsewhere on the island. We find that Airbnb revives and heightens our participants' sense of place, serving the development of community and place pride, but also conscious, overprotective place and community attachment (see Brehm *et al.*, 2004). Together, these lay the foundation for current local commitments as well as withdrawals in West Heimaey and (downtown) Heimaey, respectively. These findings underline the important duty of policy makers to shift their attention towards the needs of locals rather than tourists, such as by improving the liveability and accessibility of residential spaces and places of meaning that get lost in translation amid, particularly, professional Airbnb developments.

Moreover, our mapping reveals that despite the small geographical size and population of Heimaey, Airbnb develops in highly varying manners within distinctive spaces, each warranting individual attention. After all, our qualitative results show that withdrawals and commitments to place and community correlate to a specific Airbnb geography, that is concentrated professional Airbnb activity in the downtown area and isolated, casual, semi-professional home hostings by locals in West Heimaey, respectively. In this equation, a sense of place rather than place attachment appears to be most impactful when it comes to personal intentions to mobilize for the greater good.

In general, we find that community resilience in Airbnb is about changing sociopsychological sentiments, alterations to symbolic place meanings, and the character of places, all of which are currently imposed upon or encouraged by different types of home hostings. These findings are in line with the argument expressed by Manzo (2005), namely that concerns for identity and sense of place in terms of community resilience can outweigh those related to issues of resources, actions and social values. While this may not be a surprising finding (Buikstra *et al.*, 2010; Wilson, 2012), more significant is Airbnb's role in this equation. The current Airbnb place dynamics delineated here inform social changes in that they problematize sociopsychological sentiments, symbolic stances and functional adequacy (Berno and Ward, 2005). The cognitions in play as our participants reacted to different Airbnb activities and their resulting effects suggested, and ultimately informed, the direction people mobilized in as well as whether or not this translates into social capacity and an interest to prepare for future states of place and community. In order to arrive at these conclusions, it was crucial to avoid reducing Airbnb to one entity and labelling it a mere urban and economic concern. In fact, we argue that each Airbnb place dynamic resembles different forms of 'boom, doom, and gloom' (Lemelin *et al.*, 2010), expressed by cognitions related to sense of place, in the case of Heimaey where residents have the goal of achieving community resilience.

Undoubtedly tourism can be a vehicle for social development and a solution to remedy challenges affecting rural, peripheral and small-scale communities such as agricultural problems and population decline (Almirón *et al.*, 2006; Cànoves Valiente *et al.*, 2012; Almstedt *et al.*, 2016). In Heimaey, however, this is more likely to be achieved by motivating isolated and casual Airbnb developments over those that are concentrated and professional. Yet, both can negotiate community structure, people–place dynamics, and even heritage by inscribing new meaning, identities and histories to sites, streets and places (Prats Canals, 2006; Troncoso, 2013). After all, tourism is known to thrive from commodifying culture and identity, eventually rendering people and places as mere objects for tourist consumption (Devine, 2017). This speaks to common capitalistic and neoliberal practices that position short-term rentals as yet another tourism practice capable of carving out fear and promoting material and immaterial dispossession by economizing, touristifying and professionalizing private and public assets/spaces. Thus, in that respect, Airbnb in Heimaey is unlikely to translate into meaningful social development.

Ultimately, we suggest that tourism scholars conducting future research keep an open mind and consider the possibility that certain Airbnb developments can function as a form of spatial colonization. Such developments can have tremendous consequences for community resilience, especially for destinations with highly vulnerable social and economic structures that would be easily exploited by Airbnb.

References

Airbnb (2020a) Fast Facts. Available at: https://news.airbnb.com/fast-facts/ (accessed 20 May 2020).

Airbnb (2020b) Responsible hosting in Iceland. Available at: https://www.airbnb.com/help/article/2462/responsible-hosting-in-iceland#:~:text=According%20to%20the%20Act%2C%20home,rental%20period%20is%2030%20days (accessed 10 May 2020).

Airbnb (2020c) Vestmannaeyjar. Available at: https://www.airbnb.com.pa/s/Vestmannaeyjar/homes?refinementPaths%5B%5D=%2Fhomes&refinement_paths%5B%5D=%2Fhomes&source=structured_search_input_header&search_type=search_query (accessed 15 November 2019).

AirDNA (2019) Datafeed/API services. Available at: https://www.airdna.co/ (accessed 18 September 2019).

Albarrán Periáñez, J.D. (2020) Patrimonio territorial, políticas públicas y turismo: un análisis escalar. Los casos de Andalucía y La Sierra de Cádiz. Doctoral thesis, Complutense University, Madrid.

Albarrán Periáñez, J.D. (2021) Territorial image construction: an original perspective from the outsider. In: Gravari-Barbas, M. and Stankovic, I. (eds) *Proceedings of Heritage Hybridisations Concepts, Scales and Spaces. Una Europa PhD*. Paris 1 Panthéon-Sorbonne, Paris, France, pp. 105–108.

Almirón, A., Bertoncello, R. and Troncoso, C. (2006) Turismo, Patrimonio y Territorio. Una discusión de sus relaciones a partir de casos en Argentina. *Estudios y Perspectivas en Turismo* 15, 101–124.

Almstedt, A., Lundmark, L. and Petterson, Ö. (2016) Public spending on rural tourism in Sweden. *Fennia* 194, 18–31.

Altman, I. and Low, S.M. (1992) *Place Attachment*. Plenum Press, New York.

Álvarez, C. and San Fabián, J.L. (2012) La elección del estudio de caso en investigación educativa. *Gazeta de Antropología* 28(1), 14.

Benediktsson, K., Lund, K.A. and Huijbens, E. (2011) Inspired by eruptions? Eyjafjallajökull and Icelandic tourism. *Mobilities* 6(1), 77–84.

Berno, T. and Ward, C. (2005) Innocence abroad: a pocket guide to psychological research on tourism. *American Psychologist* 60(6), 593–600.

Bertoncello, R. (2015) Patrimonio y turismo, una relación en tensión. Aportes para el estudio en Argentina. In: Salomão, Y. (ed.) *Patrimônio Cultural Plural*. Arraes Editores Ltda, Belo Horizonte, Brasil, pp. 5–19.

Booth, A. (2015) The sharing economy, the future of jobs, and 'postcapitalism'. In: *Defence of Marxism*, 2 October 2015. Available at: http://www.marxist.com/the-sharing-economy-the-future-of-jobs-and-postcapitalism-part-one.htm (accessed 20 February 2020).

Brehm, J.M., Eisenhauer, B.W. and Krannich, R.S. (2004) Dimensions of community attachment and their relationship to well-being in the amenity-rich rural west. *Rural Sociology* 69(3), 405–429.

Brehm, J.M., Eisenhauer, B.W. and Stedman, R.C. (2013) Environmental concern: examining the role of place meaning and place attachment. *Society and Natural Resources* 26(5), 522–538.

Brown, H.S. and Vergragt, P.J. (2016) From consumerism to wellbeing: toward a cultural transition? *Journal of Cleaner Production* 132, 308–317.

Buikstra, E., Ross, H., King, C.A., Baker, P.G., Hegney, D., *et al.* (2010) The components of resilience: perceptions of an Australian rural community. *Journal of Community Psychology* 38, 975–991.

Cànoves Valiente, G., Garay Tamajón, L. and Duro Moreno, J.A. (2012) Turismo rural en España: avances y retrocesos en los últimos veinte años. *Papers de Turismo* 51, 7–21.

Devine, J.A. (2017) Colonizing space and commodifying place: tourism's violent geographies. *Journal of Sustainable Tourism* 25(5), 634–665.

Dolnicar, S. and Zare, S. (2020) COVID19 and Airbnb – disrupting the disruptor. *Annals of Tourism Research* 83: 102961.

Domínguez, A.Q. and Scarnato, A. (2017) The Barrio Chino as last frontier: the penetration of everyday tourism in the dodgy heart of the Raval. In: Gravari-Barbas, M. and Guinand, S. (eds) *Tourism and Gentrification in Contemporary Metropolises: International Perspectives*. Routledge, London.

Dredge, D. and Gyimóthy, S. (2015) The collaborative economy and tourism: critical perspectives, questionable claims and silenced voices. *Tourism Recreation Research* 40(3), 286–302.

EyjaTours (2019) Welcome to Westman Islands Vestmannaeyjar. Personal, Fun and Local Tours in Westman Islands. Available at: https://www.eyjatours.com/#:~:text=WESTMAN%20ISLANDS-,THE%20BIGGEST%20PUFFIN%20COLONY%20IN%20THE%20WORLD,the%20population%20of%20around%204300 (accessed 14 January 2020).

Folke, C., Carpenter, S.R., Walker, B., Scheffer, M., Chapin, T. and Rockström, J. (2010) Resilience thinking: integrating resilience, adaptability and transformability. *Ecology and Society* 15(4), 20.

Füller, H. and Michel, B. (2014) 'Stop being a tourist!' New dynamics of urban tourism in Berlin-Kreuzberg. *International Journal of Urban and Regional Research* 38(4), 1304–1318.

Garay, L., Morales, S. and Wilson, J. (2020) Tweeting the right to the city: digital protest and resistance surrounding the Airbnb effect. *Scandinavian Journal of Hospitality and Tourism* 20(3), 246–267.

Gual, J., Mulet, M.J. and Cueto, J. (1986) *Façanes de botigues de Palma*. Ajuntament de Palma and Gràfiques Soler, Palma de Mallorca, Spain.

Gutiérrez, J., García, J.C., Romanillos, G. and Salas-Olmedo, M.H. (2017) The eruption of Airbnb in tourist cities: comparing spatial patterns of hotels and peer-to-peer accommodation in Barcelona. *Tourism Management* 62, 278–291.

Guttentag, D. (2019) Progress on Airbnb: a literature review. *Journal of Hospitality and Tourism Technology* 10(4).

Hernández Ramírez, J. (2008) Hiperespecialización turística y desactivación del patrimonio. La gestión eclesiástica del Patio de los Naranjos de la Catedral de Sevilla. In: Hernández, J. and Díaz, E. (coord.) *Patrimonio Cultural, Turismo y Religión*. Escuela Nacional de Antropología e Historia, Mexico City, pp. 13–32.

Ioannides, D., Röslmaier, M. and van der Zee, E. (2019) Airbnb as an instigator of 'tourism bubble' expansion in Utrecht's Lombok neighbourhood. *Tourism Geographies* 21(5), 822–840.

Jóhannesson, G.T. (2015) A fish called tourism: emergent realities of tourism policy in Iceland. In: Jóhannesson, G.T., Ren, C. and van der Duim, R. (eds) *Tourism Encounters and Controversies: Ontological Politics of Tourism Development*. Ashgate, Surrey, UK, pp. 181–200.

Jorgensen, B.S. and Stedman, R.C. (2001) Sense of place as an attitude: lakeshore owners' attitudes toward their properties. *Journal of Environmental Psychology* 21(3), 233–248.

Kristjánsdóttir, H. (2016) Can the Butler's tourist area cycle of evolution be applied to find the maximum tourism level? A comparison of Norway and Iceland to other OECD countries. *Scandinavian Journal of Hospitality and Tourism* 16(1), 61–75.

Landmælingar Íslands (2019) Landmælingar. Available at: https://www.lmi.is/is (accessed 2 December 2019).

Lemelin, H., Dawson, J., Stewart, E.J., Maher, P. and Lueck, M. (2010) Last-chance tourism: the boom, doom, and gloom of visiting vanishing destinations. *Current Issues in Tourism* 13(5), 477–493.

Lyon, C. (2014) Place systems and social resilience: a framework for understanding place in social adaptation, resilience, and transformation. *Society and Natural Resources* 27(10), 1009–1023.

Mair, J. and Reischauer, G. (2017) Capturing the dynamics of the sharing economy: institutional research on the plural forms and practices of sharing economy organizations. *Technological Forecasting and Social Change* 125, 11–20.

Manzo, L.C. (2005) For better or worse: exploring multiple dimensions of place meaning. *Journal of Environmental Psychology* 25(1), 67–86.

Masterson, V.A., Stedman, R.C., Enqvist, J., Tengö, M., Giusti, M. *et al.* (2017) The contribution of sense of place to social-ecological systems research: a review and research agenda. *Ecology and Society* 22(1), 49.

Mermet, A.C. (2019) *The 'Airbnbfication' of the Icelandic Capital. Towards an Assessment of the Socio-Spatial Impacts of Airbnb in the Reykjavik Capital Area in Ferð*amálarannsóknir*. Tourism Research in Iceland, Reykjavik.

Muller, V., Coromina, L. and Galí, N. (2018) Overtourism: residents' perceptions of tourism impact as an indicator of resident social carrying capacity – case study of a Spanish heritage town. *Tourism Review* 73, 277–296.

Norris, F.H., Stevens, S.P., Pfefferbaum, B., Wyche, K.F. and Pfefferbaum, R.L. (2008) Community resilience as a metaphor, theory, set of capacities, and strategy for disaster readiness. *American Journal of Community Psychology* 41(1), 127–150.

Óladóttir, O.T. (2018) Tourism in Iceland in figures. Available at: www.ferdamalastofa.is/static/files/ferdamalastofa/talnaefni/tourism-in-iceland-2018_2.pdf (accessed 2 May 2020).

Partington, R. (2018) Tourists and tech bring resilient Iceland back from the brink. *The Guardian*, 16 June 2018. Available at: https://www.theguardian.com/world/2018/jun/16/iceland-tourists-tech-resilient-bank-from-brink-financial-crisis (accessed 10 November 2020).

Perkins, D.D., Hughey, J. and Speer, P.W. (2002) Community psychology perspectives on social capital theory and community development practice. *Journal of the Community Development Society* 33(1), 33–52.

Prats Canals, L. (2006) La mercantilización del patrimonio: entre la economía turística y las representaciones identitarias. *PH Boletín del Instituto Andaluz del Patrimonio Histórico* 58, 72–80.

Proshansky, H.M. (1978) The city and self-identity. *Environment and Behavior* 10(2), 147–169.

Raymond, C.M., Brown, G. and Weber, D. (2010) The measurement of place attachment: personal, community, and environmental connections. *Journal of Environmental Psychology* 30(4), 422–434.

Roelofsen, M. and Minca, C. (2018) The superhost. Biopolitics, home and community in the Airbnb dream-world of global hospitality. *Geoforum* 91, 170–181.

Russo, P. and Richards, G. (2016) *Reinventing the Local in Tourism*. Channel View, Bristol, UK.

Sack, R.D. (1997) *Homo Geographicus: a Framework for Action, Awareness, and Moral Concern.* Johns Hopkins University Press, Baltimore, Maryland.

Stedman, R.C. (2008) What do we 'mean' by place meanings? Implications of place meanings for managers and practitioners. In: Kruger, L.E., Hall, T.E. and Stiefel, M.C. (eds) *Understanding Concepts of Place in Recreation Research and Management*. General Technical Report PNW-GTR-744, United States Department of Agriculture Forest Service Pacific Northwest Research Station, Portland, Oregon, pp. 71–82.

Stokols, D. and Shumaker, S.A. (1981) People in places: a transactional view of settings. In: Harvey, J.H. (ed.) *Cognition, Social Behavior, and the Environment*. Lawrence Erlbaum, Hillsdale, New Jersey, pp. 441–488.

Stors, N. (2020) Constructing new urban space through Airbnb. *Tourism Geographies*, 1–24. DOI: 10.1080/14616688.2020.1750683

The Telegraph (2019) Iceland wants to curb Airbnb to cope with tourism surge. Available at: https://www.telegraph.co.uk/travel/iceland-to-restrict-airbnb-rentals-to-cope-with-tourism-surge/ (accessed 20 January 2021).

Troncoso, C.A. (2013) Espacio patrimonial, espacio de deleite visual y consumo. Transformaciones patrimoniales recientes y turismo en la ciudad de Salta (Argentina). *Espaço & Geografia* 16(2), 641–674.

UNWTO (World Tourism Organization) (2016) *Compendium of Tourism Statistics: Data 2010–2014*. UNWTO, Madrid.

UNWTO (World Tourism Organization), Centre of Expertise Leisure, Tourism & Hospitality, NHTV Breda University of Applied Sciences and NHL Stenden University of Applied Sciences (2018) *'Overtourism'? – Understanding and Managing Urban Tourism Growth beyond Perceptions, Executive Summary*. UNWTO, Madrid.

VWI (Visit Westman Islands) (2020) Welcome to Westman Islands. The best kept secret in Iceland. Available at: https://visitwestmanislands.com/ (accessed 29 December 2020).

Walker, B. and Salt, D. (2006) *Resilience Thinking: Sustaining Ecosystems and People in a Changing World.* Island Press, Washington.

Walker, B., Holling, C.S., Carpenter, S.R. and Kinzig, A. (2004) Resilience, adaptability and transformability in social-ecological systems. *Ecology and Society* 9(2), article 5.

Wilson, G. (2012) Community resilience, globalisation, and transitional pathways of decision-making. *Geoforum* 43(6), 1218–1231.

Yacuzzi, E. (2005) El estudio de caso como metodología de investigación: teoría, mecanismos causales, validación. *CEMA Working Papers: Serie Documentos de Trabajo* 296.

Zwiers, S., Markantoni, M. and Strijker, D. (2018) The role of change- and stability-oriented place attachment in rural community resilience: a case study in south-west Scotland. *Community Development Journal* 53(2), 281–300.

9 Peer-to-peer Accommodation as a Peacebuilding Tool: Community Resilience and Group Membership

Katerina Antoniou*
University of Central Lancashire, Cyprus

Abstract
Peer-to-peer (P2P) accommodation has grown exponentially in popularity over the past decade. This dynamic does not only influence the international tourism industry but carries additional sociocultural and political implications. This chapter discusses the impact of P2P accommodation in fostering positive intercultural relations and in serving as a peacebuilding tool for members of conflicting communities. In particular, the chapter examines the Couchsurfing network in Cyprus and examines its members' interactions and relations through principles of group membership. The study concludes that there is a strong group membership element in the Cyprus-based Couchsurfing community that brings together members from across the island's divide and suggests their increased resilience against divisive rhetoric.

Keywords: P2P accommodation, community resilience, peacebuilding, group membership, Couchsurfing, conflict resolution

Highlights

- Community resilience can act as a condition and not as an end goal in processes of reconciliation.
- Community resilience is stronger in groups with a strong and positive sense of group membership.
- Group membership is dynamic and individuals are members of multiple social groups at the same time, prioritizing the groups with a stronger sense of group membership.
- The Couchsurfing platform has fostered the development of a vibrant intercommunal community of hosts in Cyprus, revealing its ability to create strong bonds across members of divided societies.
- The Couchsurfing platform and other P2P-hospitality platforms show significant potential to create inclusive communitas for people living in conflict, and therefore can be further considered for their ability to contribute to peace.

Introduction

Interpersonal relations across communities in conflict have long been at the forefront of conflict

*KAAntoniou@uclan.ac.uk

 Peer-to-peer Accommodation and Community Resilience (eds Anna Farmaki, Dimitri Ioannides and Stella Kladou)
DOI: 10.1079/9781789246605.0009

resolution studies (Fisher, 1997; Oakes, 2004; Maoz, 2011). Achieving sustainable peace between former rivals is almost exclusively measured by the disputants' ability to collaborate, trust one another, and achieve their joint transition – politically, constitutionally and socially. To this end, effective conflict resolution relies on two parameters: (i) a sense of belonging to a common community – in other words, an inclusive and prevalent group membership; and (ii) collective – or community – resilience against the challenges of post-conflict transformation.

Community resilience, as defined by Magis (2010, p. 401), is the 'existence, development, and engagement of community resources by community members to thrive in an environment characterized by change, uncertainty, unpredictability, and surprise'. This feature is particularly critical for societies recovering from conflict, as they are undergoing – or are expected to experience – a post-conflict transitional period of change, uncertainty and unpredictability. Community resilience can thus be a key factor in determining the ability of a reconciling society to successfully develop sustainable forms of peaceful coexistence.

The association between community resilience and group membership is better understood by identifying models of social structure that effectively promote resilience and sustainability. Turner (1969) distinguishes the Latin term **communitas** from the term community in order to emphasize the former's connotation for stronger social bonds. Wang (1999) presents the concept of **touristic communitas**, referring to the interpersonal ties formed during travelling. The author speaks of genuine emotional bonds formed in a way that generates interpersonal authenticity. Říhová *et al.* (2013) introduce the term **temporary communitas**, highlighting that some communities create strong bonds within a confined temporal framework. Molz (2014) breaks down the categories of tourist activity that can generate stronger social bonds even further, identifying network hospitality as an emerging form of authentic interpersonal connections. According to the author, network hospitality refers to the use of networks and platforms enabling peer-to-peer (P2P) hospitality, with particular emphasis on non-profit networks such as Couchsurfing.

In this chapter, I examine social bonds within the Couchsurfing community, specifically in reference to the ability of the Couchsurfing community to develop more resilient and sustainable communities. Couchsurfing, a network of free hospitality exchange (Molz, 2012), is seen to achieve the socio-psychological transformation of those involved (Decrop *et al.*, 2018). Members of the community have testified to achieve a sense of spiritual fulfilment through their Couchsurfing experience (Molz, 2014). The chapter discusses the transformative power of the Couchsurfing network by examining the Couchsurfing group membership within communities in conflict, and by assessing whether its transformative powers can contribute to the creation of inclusive social bonds and increased community resilience.

The chapter addresses its objectives by focusing on the case study of Cyprus, a geographically partitioned island state experiencing a protracted ethnic conflict between its Greek Cypriot and Turkish Cypriot communities. Cyprus features an active Couchsurfing community that can be examined in reference to the country's prolonged ethnic conflict. In particular, the island's Couchsurfing community has developed in an intercommunal character, engaging members from both sides of the island's divide, indicating the community's transcendence beyond physical and sociopolitical barriers. Cyprus, therefore, provides a useful example for assessing the contribution of network hospitality and P2P accommodation to community resilience across communities in conflict, and to that end, its ability to foster sustainable peace in processes of post-conflict transition and reconciliation.

Community Resilience and Group Membership

Membership in a community is expressed through collective experiences, which provide unique forms of shared cognition to the members of that group (Tindale *et al.*, 2004). Worchel and Coutant (2004) acknowledge the temporal nature of group membership and introduce the process of group development through four stages: (i) group identification; (ii) group productivity; (iii) individuation; and (iv) decay. Group identity and group membership are most salient at the group identification and group productivity stages, whereas personal identity surpasses the salience of group identity during the stages of individuation and decay (Worchel and Coutant, 2004).

Group membership and shared experiences are relevant for understanding individual attitudes towards the other members of one's community, and against individuals outside the collective unit. According to extant research, the categorization of individuals into social groups is likely to promote hostility and discrimination between groups (Brewer and Gaertner, 2004; Oakes, 2004; Rothbart and Korostelina, 2006). Nevertheless, the deconstruction of in-group boundaries is possible, particularly through narrative that allows individuals to see themselves as members of wider, more inclusive units (Brewer and Gaertner, 2004). Antoniou (2019) supports that threatened group identities can be de-securitized when rival groups in a conflict develop a common identity with a stronger group membership, one that offers transformative shared experiences.

Collective experiences can transform how social identities are perceived and generate alliance to wider groups, even in the presence of ongoing ethnic conflict. According to Allport's (1954) Contact Hypothesis, this can be achieved when members of alien or antagonistic groups engage in meaningful interaction, under a set of optimal conditions. When the conditions of equal status, authority support, shared goal and intergroup collaboration are met, the interaction can yield positive encounters, and potentially eliminate prejudice or discrimination between the groups involved. This is reiterated by McMillan and Chavis (1986), who argue that the Contact Hypothesis directly links to the principle of shared emotional connection within a community.

To understand the transformative capacities of collective experiences, it is important to identify the parameters affecting a group's dynamics and how each member understands their position within the group. Cameron (2004) introduces a three-factor model that assesses group membership from the perspective of social identity. The model includes **centrality**, **in-group affect** and **in-group ties** as the three principal factors in assessing group membership prioritization. Centrality refers to the importance of a group for the individual, based on the frequency in which it comes to mind. In-group affect indicates the emotion that group membership generates, whether it is pride, joy or collective guilt. The third component of Cameron's model is in-group ties, or the emotional connections formed between a group's members through their shared experiences. According to Tindale *et al.* (2004), shared cognition in small groups endorses group cohesiveness, with the latter being articulated as a key component of social identity and group membership.

As individuals experience group membership multiplicity and affiliate with multiple social groups at any given time, this model provides an insight into the parameters that make one group membership more important than another. Cameron's model enables us to examine the social categorization processes individuals undertake, and whether their membership to wider, more inclusive groups makes exclusive social identities less relevant. This perspective, as discussed below, creates a direct association between group membership and community resilience, a relationship worth examining for communities experiencing conflict.

Community resilience through inclusive group membership

Pandey (2014) presents community resilience as an end goal in a conflict's roadmap to reconciliation. Juncos (2018) challenges this position and suggests that community resilience should not be treated as a goal in peacebuilding, but as an approach to resolving the conflict more effectively. As reiterated by Juncos and Joseph (2020), the role of resilience in international peacebuilding interventions has considerably increased. Today, community resilience can lead to sufficient community responses against conflict backlashes. Based on evidence from Tanzania, increased community resilience serves to both prevent and counter violent conflict (Walwa, 2018). It can thus provide a firm ground for sustainable peace, or as Johnson *et al.* (2012, p. 48) define it, to make peace last by 'creating and maintaining effective cooperative systems among diverse groups despite different interests'.

Community resilience is, therefore, a key factor to consider in the process of creating more inclusive communities and a common group membership for former rivals. Transformative experiences that are shared by members of rival communities can provide the grounds for a common, inclusive community membership, and contribute to processes of reconciliation and peace.

Inclusive communities can be significantly more resilient in the presence of strong social bonds. According to Beavitt (2012), communities have a symbolic significance to their members. A community's positive symbolism and resilience relies on its ability to establish liminal spaces that enable communities to become communitas and establish stronger bonds across their members (Beavitt, 2012). Communitas allow their members to engage in transformative shared experiences, what Turner (1969) refers to as a **rite of passage**. Communitas can, therefore, be understood as resilient models of communal structures, characterized by strong group membership that can lead to collective transformation. The section below delves further into the existence of tourism-generated communitas which, through strong group membership and transformative experiences, can contribute to community (re)formation and community resilience.

Tourism, Peace and Group Membership

Travel and tourist activity, both domestic and international, has frequently been heralded as an activity closely linked to reconciliation and peace, with scholars debating whether tourism can have causal effects over the ability of rival groups to reconcile (Salazar, 2006; Moufakkir and Kelly, 2010; Pratt and Liu, 2016; Farmaki, 2017). Anastasopoulos (1992) and Salazar (2006) have questioned the ability of tourism to engender peace, while Farmaki (2017) suggests that the relationship between tourism and peace remains inconclusive. In a more recent account, Tavares da Silva *et al.* (2020) reinforce the notion of tourism as a contributor to peace, by supporting that tourism acts as a bottom-up approach to the transformation of conflict by confronting barriers to reconciliation and ensuring the sustainability of post-conflict transition towards peace.

Despite the extensive scholarly attention paid to the relationship between tourism and peace, group membership and social categorization are yet to be employed in the tourism-peace discussion. Social categorization and re-categorization is central to conflict resolution research, and therefore principal to scholarly discussions on peace. Yet, there is an evident gap when discussing the contribution of tourism to peace without addressing the ability of tourism to contribute to processes of social re-categorization. This chapter addresses this gap by applying the perspective of social categorization and group membership to a specific tourist activity: P2P accommodation.

The role of P2P-hospitality platforms

P2P accommodation has grown exponentially through the availability of online platforms offering, as Lauterbach *et al.* (2009) put it, hospitality services. These platforms establish well-networked online communities that rely on a reputation system (Lauterbach *et al.*, 2009) ensuring the credibility and reliability of their users and service providers. The authors mention that there is a wide variety of transnational online hospitality communities, including Servas International, Global Freeloaders, Hospitality Club, and Couchsurfing. Servas International clarifies through their branding and activity that its members share a common set of values as supporters of peace internationally. The community particularly uses the slogan *Peace and Understanding through Travel and Hosting*, while they also closely collaborate with the United Nations (UN) on peace-related initiatives (Servas.org, 2020). Servas International poses a valid case study for examining peace through tourism from the perspective of tourist communities and P2P-hospitality service provision. Nevertheless, the Couchsurfing community provides an even more compelling case study with insights of 'sociological and cultural significance' (Molz, 2012, p. 215), as it is 'by far the largest and most popular online hospitality service, with more than one million members representing 230 countries' (Lauterbach *et al.*, 2009, p. 346).

The ability of P2P-accommodation platforms to foster inclusive communities across national and ethnic boundaries enables some of these hospitality networks to achieve strong social bonds across their members. In reference to community resilience, hospitality networks with strong social bonds and characteristics of communitas could serve as more resilient communities, whose membership overshadows the exclusivity of divisive, ethnic identity, and allows its members to come together under a new, common identity.

Couchsurfing has been identified as a type of tourist activity that is socially transformative (Molz, 2012; Decrop *et al.*, 2018). It is also a community within which the reciprocity of hospitality service provision among the community's members is central to the platform's operations (Lauterbach *et al.*, 2009). This observation suggests the ability of Couchsurfing to engage its members in transformative experiences and enable their identity re-categorization. The Couchsufring network is therefore a P2P-accommodation platform worth examining within the context of conflict resolution, and with regards to group membership and community resilience.

Couchsurfing in Cyprus

The Couchsurfing community operates through an online network of exchanges and virtual interactions, before these materialize through hosting, travelling and other shared social activities. The virtual nature of the community enables it to bring the community's members together without them conforming to a common physical location. This enables online hospitality networks and the Couchsurfing community in particular, to connect people across national borders, as well as people across division barriers, buffer zones and societies in conflict. The fact that Couchsurfing is a platform of free hospitality exchange and not a paid service provider further enables it to surpass the limitations of economic transactions between communities in conflict.

As a result, the ability of Couchsurfing to establish communities of free hospitality exchange makes it an appropriate platform for connecting people of geographical proximity without highlighting their ethnic or national affiliations. This is particularly significant for the case of Cyprus, which hosts a population that is in close geographical proximity but is otherwise ethnically and politically partitioned. This Mediterranean island features a discourse of civil unrest between its ethnically Greek majority and its ethnically Turkish minority, which commenced soon after its independence in 1960 and escalated in armed conflict with interstate dimensions. Turkey performed a military intervention in 1974, a development that led to the geographical partition of the two communities consolidated with an island-wide, UN-administered Buffer Zone (Farmaki and Antoniou, 2017).

Nearly five decades later, the Cyprus conflict remains unresolved, with the two communities functioning under separate authorities – the Greek-speaking south is under the internationally recognized Republic of Cyprus, while the Turkish-speaking north falls under the self-proclaimed 'Turkish Republic of Northern Cyprus'. The two authorities refrain from any diplomatic collaboration, and as such tourist activity is de facto operated separately for the two partitioned communities. The Couchsurfing platform has enabled members from both communities in Cyprus to jointly create a local Couchsurfing chapter that includes Couchsurfers from across the island's divide. The Couchsurfing website features approximately 11,000 Cyprus-based hosts, including both locals of Greek and Turkish ethnic backgrounds, as well as residents of Cyprus of other origins and nationalities (Couchsurfing.com, 2020).

Methodology

In this chapter, I adopt an exploratory qualitative research approach towards the case study of Couchsurfing in Cyprus, both by reviewing Couchsurfing activity in Cyprus, as well as by incorporating feedback from Couchsurfers locally and regionally.

A set of primary data was collected through one-to-one interviews that featured a brief questionnaire and informal conversation with follow-up questions. Interview participants were identified through social media platforms and personal communication. The majority of participants were local hosts in Cyprus. Other interview participants include foreign Couchsurfers outside of Cyprus, whose testaments help to provide a wider framework of Couchsurfer activity beyond the case study of Cyprus.

Additional complementary data includes information from secondary sources, such as the Couchsurfing website, Couchsurfing social media pages and Couchsurfing-related blogs. I have treated data collected through interviews confidentially and in respect to data protection protocols. All participants have been anonymized, and their input is presented through pseudonyms that reflect their gender and ethnicity. More information on participant profiles and collected data is included in Table 9.1.

The following secondary group sources were used:

- official Couchsurfing website:
  - Coachsurfing.com (2020) Available at: www.couchsurfing.com (accessed 2 July 2020);
  - https://www.couchsurfing.com/places/cyprus (accessed 2 July 2020);
- affiliated blogs:
  - https://skyetravels.com/couchsurfing-in-cyprus/ (accessed 2 July 2020);
  - https://blog.couchsurfing.com/ (accessed 9 September 2020);
  - https://blog.couchsurfing.com/10-questions-with-a-cs-ambassador-pandelis/ (accessed 9 September 2020);
  - https://blog.couchsurfing.com/a-couchsurfers-guide-to-nicosia/ (accessed 9 September 2020);
  - https://www.nomadicmatt.com/travel-blogs/how-to-crush-it-on-couchsurfing/ (accessed 9 September 2020);
- social media pages on Couchsurfing (Facebook, Twitter, Instagram): https://www.facebook.com/Cyprus-Couchsurfing-Camping-Site-1228437120643513/ (accessed 9 September 2020); and
- YouTube.

Findings from the study's data collection process have been analysed in reference to Cameron's (2004) group membership model and its three factors of: (i) centrality; (ii) in-group affect; and (iii) in-group ties. This approach provides an insight to social categorization processes within the community of Couchsurfers and assesses the extent to which the community endorses inclusive group membership, and potentially contributes to reinforcing resilience for communities in conflict.

Findings

Hosting and getting together

Couchsurfing has formed a vibrant community in Cyprus that hosts international visitors and brings local hosts together on a frequent basis, through get-togethers for coffee or drinks. The Couchsurfing platform is the first point of contact for potential visitors, who can ask for guidance and travel advice before finalizing their trip. Local hosts offer accommodation, but the option of camping is also available.

The Greek and Turkish Cypriot communities of Cyprus have been geographically partitioned since 1974, a reality that bears a number of restrictions to movement and transportation for the island's visitors and residents. The two communities are separated by the UN-administered Buffer

Table 9.1. Profile of interviewees.

Name	Gender	Age (years)	Ethnicity	Couchsurfing activity
Marios	M	30–39	Greek Cypriot	Active
Meltem	F	30–39	Turkish Cypriot	Active
Stelios	M	30–39	Greek Cypriot	Affiliated to active member
Deniz	F	30–39	Turkish (Cyprus resident)	Formerly active
Grace	F	30–39	French (Cyprus resident)	Formerly active
Ayshe	F	30–39	Turkish (non-Cyprus resident)	Formerly active
Donald	M	30–39	American (non-Cyprus resident)	Active
Amina	F	30–39	Bosnian (non-Cyprus resident)	Formerly active

Zone that runs across the island and separates its northern and southern territories. Since 2003, both visitors and members of each community have been able to cross the Buffer Zone by providing a passport or national ID issued by the Republic of Cyprus (RoC citizenship has been issued to both Greek and Turkish Cypriots). Nevertheless, accessing the southern part of the island is not possible for Turkish nationals or Turkish Cypriots that do not hold a Republic of Cyprus ID. Access is also restricted to visitors or non-permanent residents without valid documentation.

Despite challenges to accessibility, the Couchsurfing community has been able to bring together members residing on either side of the Buffer Zone. A contributing factor was the launch of a multipurpose venue and café called the *Home for Cooperation*, which was established in Nicosia in 2012. The Home for Cooperation is situated within an accessible part of the UN-administered Buffer Zone between the two communities and provides Couchsurfer hosts from both communities with a venue where they can meet without crossing into the other community. Intercommunal meetings have been recorded for the past decade, and many have been taking place within and close to the Nicosia Buffer Zone (Pandelis, 2019, 2020).

Social gatherings over coffee, food and drinks appear to be a popular activity for the Cyprus Couchsurfing community. According to the testaments of local hosts, these individuals more frequently go out in the company of other hosts rather than host visitors and provide hospitality services. While hosting a visitor might take place once or twice annually, social activities with local Couchsurfers can take place as frequently as once or twice a month. This experience was shared both by active Couchsurfer Marios (2020, personal communication) and formerly active Deniz, who spoke about her past engagement with the platform:

> Unfortunately, I am not involved in Couchsurfing anymore. Back in 2013, I remember organizing a number of Couchsurfing gatherings. They were like social gatherings; people, foreigners and both communities in Cyprus coming together and chatting ... I also used the café at the Home for Cooperation [as a gathering spot], because it's in the Buffer Zone so everyone can come and also, [it offers] a nice environment. [You could never tell] how many people will come, because you never know. One time I organized we were around 30 people, another one only five; so, it really changes.
>
> (Deniz, 2020, personal communication)

Engagement in intercommunal group activities was also reiterated by Turkish Cypriot Couchsurfer, Meltem:

> I am on Couchsurfing since 2009, and I hooked my [Greek Cypriot] husband too. It would take a [great] amount of time to try and count all the Couchsurfing activities I have been involved in. I feel that we would be the exception [to your findings] since I have been hosting on both sides of the island! You can see our activities on Facebook if you join. It's a good experience, hop in!
>
> Couchsurfing has its own culture, and once you understand the way it works, no matter from which country and/or ethnicity, a Couchsurfer is a Couchsurfer. [In my experience] ethnicities and destinations didn't matter, as long as the surfers were true.
>
> (Meltem, 2020, personal communication)

Couchsurfing membership and motivation to join

Meltem's sentiments on a common Couchsurfer identity and sense of community were echoed by Grace, who spoke of long-term friendships, life-changing experiences, but also a lost sense of community after the platform became popular:

> My husband was also a member of Couchsurfing even though we did not meet through Couchsurfing. He is Turkish Cypriot, living in the Republic of Cyprus. Some of our long friends to date are people he met and introduced me to via the Couchsurfing meetings. These people are from different ethnic groups (Turkish Cypriots, Greek Cypriots, Armenian Cypriots, Maronites, expats living in Cyprus, etc.).
>
> [Couchsurfing] really opened my eyes to the diversity of societies and the need to explore this diversity. It made me more empathetic towards a community when it is a minority, made me more aware of some cultural/political sensitivities and ways to respect them. It also made me less accepting of stereotypic statements in e.g. newspapers or other media about that community.
>
> Couchsurfing maybe played a role in me getting degrees in international relations, intercultural

> communication and then accepting an internship in Cyprus with a bicommunal programme.
>
> At some point, receiving too many requests from people to get hosted, I started being very selective so as to meet people who understood Couchsurfing was not only a [means] to find free accommodation but was about meeting people. Actually, when Couchsurfing got so popular, I started losing interest – it became a mainstream thing and I was getting upset by the opportunistic requests for hosting and by the fact that the spirit that was there at the beginning of the community was somehow lost. (Grace, 2020, personal communication)

A lost sense of community was also highlighted by a US-based, American Couchsurfer, Donald:

> In general, I love using Couchsurfing, but it can be different based on gender. There are a lot of creepers on it that have ruined the idea and concept of what it means. My girlfriend visited when we lived in Moldova but she did Airbnb. (Donald, 2020, personal communication)

Gender, risks and limitations

Donald raises the issue of a weak sense of community, and at the same time the consequences of losing trust in the platform and dealing with increased risk, particularly in reference to gender-based violence. The latter is also stressed by a Turkish female Couchsurfer, Ayshe: 'I only hosted female Couchsurfers and stayed in female hosts' (Ayshe, 2020, personal communication).

Participant testimonies overwhelmingly refer to solo travelling when using and offering hospitality services through Couchsurfing. Ayshe, who only actively engaged with other female Couchsurfers, also mentioned that her engagement with Couchsurfing stopped after getting married, having a baby and having other priorities and time commitments in her life.

In addition to perceived gender-based risk and precautions, a blogger's journal entry on 'Couchsurfing in Cyprus' raises the risk of sexual assault within the Cyprus context, with the blogger naming a local host to warn guests that he has a reputation of assault (SkyeTravels.com, 2018). The blogger also mentions the risk of host unreliability, as he experienced a sequence of no responses or delayed communication while being on the island, and as a result he had to secure alternative options of last-minute accommodation. In a conversation I had with Stelios (2020, personal communication) – who is not a Couchsurfer but is friends with a local host – he mentioned an incident when he hosted a Couchsurfer for a night, because his host friend did not have enough space, reaffirming a risk of unreliability in the hospitality exchanges communicated through the platform.

Discussion and Future Implications

In addition to the aforementioned risks, the data collected presents an interesting discrepancy between visitor–host interactions and the local host community in Cyprus. While the active provision and reception of hospitality services through the Couchsurfing platform is seen to be primarily undertaken by young and single individuals, the group activities of local hosts in Cyprus are seen to surpass that demographic, including married couples and former Couchsurfers. This indicates that there might be certain demographic limitations to one's active engagement with the Couchsurfing community as a platform for hospitality service exchanges, whereas this limitation is not so evident in local host interactions and shared experiences. Ayshe's temporary and gender-confined engagement with Couchsurfing comes in a striking contradiction to the Cyprus host community's activities, which have fostered long-term friendships. The Cyprus-based hosts also feature a more diverse demographic, with a mixture of genders, religions and ethnicities across participants, who are open to inviting trusted affiliates to the hosts, such as friends or partners.

The findings suggest that the Couchsurfing experience has the capacity to be transformational, not only with regards to temporary visitor–host encounters, but also for the hosts who see their Couchsurfing affiliation as an identity they share with other local hosts. Meltem and Grace identify a common sense of community that suggests the existence of an inclusive Couchsurfing group identity and membership. To this end, and in addition to the ability of Couchsurfing to establish temporary communitas across hosts and visitors, the Cyprus example features meaningful, inclusive and long-term interactions between local hosts from both of the island's divided communities. This observation implies that an inclusive Couchsurfing membership is prioritized over the divisive ethnic identities of the Couchsurfers engaging in intercommunal encounters and shared activities.

The Couchsurfing group membership shared by the intercommunal group of Cyprus' hosts can be further understood through Cameron's (2004) three factor membership model, and specifically the factors of centrality, in-group affect and in-group ties. In-group ties are evidently strong, since the community members do not only engage in shared activities beyond their engagements as hosts, but they do so on a frequent basis that suggests the long-term affinity and emotional connection of these members.

The frequent and voluntary social interactions and informal gatherings between the Couchsurfer hosts are an indicator of strongly positive emotions generated by the group, and therefore an indicator of strongly positive in-group affect. Participant testaments were overwhelmingly positive in reference to their capacity as Couchsurfer hosts, while participants were delighted to contribute to the study and discuss their Couchsurfing affiliation. This was also true for formerly active members, who often referred to their Couchsurfing identity as an ongoing one and spoke fondly of the experiences they shared with other 'surfers'. In reference to centrality, the frequency of host-to-host interactions and overall engagement is significantly higher than of host–visitor encounters. This signifies high group centrality for the local hosts, even in cases of hosts who no longer provide hospitality to visitor Couchsurfers.

In reference to Worchel and Coutant's (2004) stages of group development, the Couchsurfer host community of Cyprus appears at a stage of group productivity, indicating the importance and value of the Couchsurfing identity to its members. The presence of strong and non-temporary social bonds among Cyprus-based Couchsurfing hosts satisfies the characteristics of a communitas, a form of community that is associated with higher levels of community resilience.

Findings propose that members of the Couchsurfing host community of Cyprus, who were able to overcome the exclusivity of their rival ethnic identities and form strong social bonds with one another, are more resilient citizens because of their transformative Couchsurfing experiences. As Grace highlights, her shared experiences with members of other cultures and minority communities made her more empathetic of others and less accepting of divisive rhetoric. In light of the inclusive Couchsurfing membership being prioritized over exclusive ethnic identities, local Couchsurfers portray increased resilience by developing shared cognition and strong social bonds on an intercommunal level. What is more, local Couchsurfers have created and maintained 'effective cooperative systems among diverse groups despite different interests', which Johnson *et al.* (2012, p. 48) see as the basis of sustainable peace.

In an attempt to inform conflict resolution and peacebuilding processes, I have examined Couchsurfing with reference to community resilience and group membership. Through the case study of Cyprus, the P2P-hospitality network confirmed its ability to both increase community resilience across its members, as well as achieve their re-categorization into a wider, more inclusive in-group. This is achieved through the formation of a Couchsurfer communitas among local hosts that appears to maintain its relevance and centrality in the long term. Although the Couchsurfing platform was thus far considered a touristic communitas, offering temporary or one-off transformational experiences, findings from Cyprus suggest that it can also foster lasting group membership, shared across lines of ethnic division.

The ability of the Couchsurfing platform to provide common ground for members of divided communities is a significant and promising discovery, as it directly informs peacebuilding processes. Positive intergroup contact and group re-categorization lie at the forefront of international peacebuilding interventions, and the Couchsurfing platform offers an organic structure that directly achieves these objectives. The P2P network brings together its members as equal partners, sharing their common desire to travel and engage in authentic interactions with one another. These characteristics satisfy Allport's (1954) conditions for positive intergroup contact, further reaffirming the platform's ability to directly accommodate peacebuilding and conflict transformation.

Having in mind the ability of Couchsurfing to provide a non-temporary shared identity across conflict rivals, a key recommendation of this chapter is to examine this peacebuilding capacity of P2P-hospitality networks even further. Is this an isolated example of the Cyprus-based host community? Are there additional P2P-hospitality structures with peacebuilding capacities analogous to the Couchsurfing example? Do other hospitality platforms operate across conflict lines? Expanding the relevance and applicability of this chapter's research findings can directly inform peacebuilding professionals and policy makers.

Conclusion

This study has revealed that platforms of P2P accommodation, such as the examined case of Couchsurfing, can often go beyond the provision of hospitality services and foster the organic creation of resilient communities with a strong group membership. The case of a vibrant and sustainable intercommunal community of local Couchsurfers in Cyprus exposes the ability of the Couchsurfing platform to serve as an inclusive group identity and, under the conditions of an intractable conflict, to directly contribute to reconciliation. In light of these findings, the Couchsurfing platform can be considered an organic peacebuilding tool that has the potential to bring members of divided societies together from a bottom-up approach.

The study's findings are subject to a number of limitations, such as the temporal and geographical constraints of the audience examined. The chapter provides a set of preliminary observations that inform tourism and conflict resolution literature. The findings provide directions for future research that can further expand the scope, applicability and validity of what is observed through the Cyprus case study. Recommendations for future research can be derived from the examination of Couchsurfing and other P2P-accommodation platforms in societies experiencing conflict, and their ability to generate strong social bonds and inclusive memberships across rival communities.

Cyprus provides an example of Couchsurfing hosts coming together despite the political, ethnic and geographical division they experience, and forming a communitas of strong social bonds that transcends social conventions. Looking beyond Cyprus, these findings directly inform divided societies across the world, whose citizens can utilize P2P-accommodation platforms to establish meaningful and long-term contact with hosts from across the division line. In cases where contact across division lines remains a taboo or entails legal or political restrictions, P2P-platform providers can ensure the safe and legal contact of their members by acknowledging such restrictions and ensuring the avoidance of controversies with local authorities. Intractable conflicts and post-settlement societies such as Israel and Palestine, Sri Lanka, Crimea and Northern Ireland – as well as racially segregated societies such as the USA – could benefit from the examination of P2P networking as a mechanism that enhances sustainable peace.

In this chapter I have established a positive relationship between the Couchsurfing platform and the resilience of communities in conflict. My findings pave the way for future research to examine the contribution of P2P-hospitality platforms to processes of reconciliation more widely, and the conditions in which they can become effective structures for peacebuilding and conflict transformation.

References

Allport, G. (1954) *The Nature of Prejudice*. Addison-Wesley Publishing, Cambridge, Massachusetts.

Anastasopoulos, P.G. (1992) Tourism and attitude change: Greek tourists visiting Turkey. *Annals of Tourism Research* 19(4), 629–642.

Antoniou, K. (2019) Beyond the speech act: contact, desecuritization, and peacebuilding in Cyprus. In: Butler, M.J. (ed.) *Securitization Revisited: Contemporary Applications and Insights*. Routledge, London, pp. 168–193.

Beavitt, R. (2012) The demands of liminality: community, communitas, and reflexivity. PhD thesis, Murdoch University, Perth, Australia.

Brewer, M.B. and Gaertner, S.L. (2004) Toward reduction of prejudice: intergroup contact and social categorization. In: Brown, R. and Gaertner, S.L. (eds) *Blackwell Handbook of Social Psychology: Intergroup Processes*. Blackwell, Oxford, pp. 451–472.
Cameron, J.E. (2004) A three-factor model of social identity. *Self and Identity* 3(3): 239–262.
Couchsurfing.com (2020) Available at: www.couchsurfing.com (accessed 1 July 2020).
Decrop, A., Del Chiappa, G., Mallargé, J. and Zidda, P. (2018) 'Couchsurfing has made me a better person and the world a better place': the transformative power of collaborative tourism experiences. *Journal of Travel & Tourism Marketing* 35(1), 57–72.
Farmaki, A. (2017) The tourism and peace nexus. *Tourism Management* 59, 528–540.
Farmaki, A. and Antoniou, K. (2017) Politicising dark tourism sites: evidence from Cyprus. *Worldwide Hospitality and Tourism Themes* 9(2), 175–186.
Fisher, R.J. (1997) *Interactive Conflict Resolution*. Syracuse University Press, New York.
Johnson, D.W., Johnson, R.T. and Tjosvold, D. (2012) Effective cooperation, the foundation of sustainable peace. In: Coleman, P.T. (ed.) *Psychological Components of Sustainable Peace*. Springer, New York, pp. 15–53.
Juncos, A.E. (2018) Resilience in peacebuilding: contesting uncertainty, ambiguity, and complexity. *Contemporary Security Policy* 39(4), 559–574.
Juncos, A.E. and Joseph, J. (2020) Resilient peace: exploring the theory and practice of resilience in peacebuilding interventions. *Journal of Intervention and Statebuilding* 14(3), 289–302.
Lauterbach, D., Truong, H., Shah, T. and Adamic, L. (2009) Surfing a web of trust: reputation and reciprocity on couchsurfing.com. In: *Proceedings of the 12th IEEE International Conference on Computational Science and Engineering,* 29–31 August 2009, Vancouver, Canada, Vol. 4. Institute of Electrical and Electronics Engineers (IEEE) Computer Society, New York, pp. 346–353.
Magis, K. (2010) Community resilience: an indicator of social sustainability. *Society and Natural Resources* 23(5), 401–416.
Maoz, I. (2011) Does contact work in protracted asymmetrical conflict? Appraising 20 years of reconciliation-aimed encounters between Israeli Jews and Palestinians. *Journal of Peace Research* 48(1), 115–125.
McMillan, D.W. and Chavis, D.M. (1986) Sense of community: a definition and theory. *Journal of Community Psychology* 14(1), 6–23.
Molz, J.G. (2012) Couchsurfing and network hospitality: 'It's not just about the furniture'. *Hospitality and Society* 1(3), 215–225.
Molz, J.G. (2014) Toward a network hospitality. *First Monday* 19(3). Available at: https://www.academia.edu/6343762/Toward_a_network_hospitality (accessed 2 February 2021).
Moufakkir, O. and Kelly, I. (eds) (2010) *Tourism, Progress, and Peace*. CAB International, Wallingford, UK.
Oakes, P. (2004) The root of all evil in intergroup relations? Unearthing the categorization process. In: Brown, R. and Gaertner, S.L. (eds) *Blackwell Handbook of Social Psychology: Intergroup Processes.* Blackwell, Oxford, pp. 102–119.
Pandelis (2019) A Couchsurfer's Guide to... Nicosia. *Couchsurfing Blog*, 19 September. Available at: https://blog.couchsurfing.com/a-couchsurfers-guide-to-nicosia/ (accessed 20 September 2020).
Pandelis (2020) 10 Questions with a CS Ambassador: Pandelis. *Couchsurfing Blog*, 11 May. Available at: https://blog.couchsurfing.com/10-questions-with-a-cs-ambassador-pandelis/ (accessed 20 September 2020).
Pandey, V. (2014) Community policing for conflict resolution and community resilience. *International Journal of Social Work and Human Services Practice* 2(6), 229–233.
Pratt, S. and Liu, A. (2016) Does tourism really lead to peace? A global view. *International Journal of Tourism Research* 18(1), 82–90.
Říhová, I., Buhalis, D., Moital, M. and Gouthro, M. (2013) Social layers of customer-to-customer value co-creation. *Journal of Service Management* 24(5), 553–566. DOI: 10.1108/JOSM-04-2013-0092
Rothbart, D. and Korostelina, K.V. (2006) *Identity, Morality and Threat: Studies in Violent Conflict*. Rowman & Littlefield, Lanham, Maryland.
Salazar, N.B. (2006) Building a 'culture of peace' through tourism: reflexive and analytical notes and queries. *Universitas Humanística* 62, 319–336.
Servas.org (2020) Servas International. *Peace and Understanding through Travel and Hosting*. Available at: www.servas.org (accessed 21 July 2020).
SkyeTravels.com (2018) Journal entry: My SNAFU with Couchsurfing in Cyprus. *Skye Travels*, 15 December. Available at: https://skyetravels.com/couchsurfing-in-cyprus/ (accessed 27 September 2020).

Tavares da Silva, J., Breda, Z. and Carbone, F. (2020) *Role and Impact of Tourism in Peacebuilding and Conflict Transformation*. IGI Global, Hershey, Pennsylvania.
Tindale, R.S., Meisenhelder, H.M., Dykema-Engblade, A.A. and Hogg, M.A. (2004) Shared cognition in small groups. In: Brewer, M.B. and Hewstone, M. (eds) *Blackwell Handbook of Social Psychology: Group Processes*. Blackwell, Oxford, pp. 1–30.
Turner, V. (1969) Liminality and communitas. In: *The Ritual Process: Structure and Anti-structure*. Aldine Transaction, Piscataway, New Jersey, pp. 125–130.
Walwa, W.J. (2018) Understanding community resilience amidst rising violent conflicts in Tanzania. *The African Review* 45, 124–142.
Wang, N. (1999) Rethinking authenticity in tourism experience. *Annals of Tourism Research* 26(2), 349–370.
Worchel, S. and Coutant, D. (2004) It takes two to tango: relating group identity to individual identity within the framework of group development. In: Brewer, M.B. and Hewstone, M. (eds) *Blackwell Handbook of Social Psychology: Group Processes*. Blackwell, Oxford, pp. 461–481.

10 'Not in My Stairway': How Do Neighbours Cope with Peer-to-peer Rentals in Housing Cooperatives?

Szilvia Gyimóthy*
Copenhagen Business School, Denmark

Abstract

This chapter approaches resilience as a problem-solving practice in neighbourhood communities, by discussing how peer-to-peer (P2P)-accommodation platform rentals have been embraced in a Nordic context. I review the evolution and cultural history of Danish housing cooperative movements as mundane hotbeds of participatory democracy and argue that everyday Nordic practices of cohabitation are significant for building resilience. Departing from the notion of the community as performatively constituted, I provide a brief illustration of how neighbourhoods are enacted in residents' images or notions of cohesion as well as of shared values, practices and spaces. Based on empirical illustrations from disputes concerning P2P rentals and construction of private balconies in Copenhagen, my position is that neighbourhood conflicts challenge, but may also enhance, societal cohesion, because they revive discussions about the rights and duties of community members.

Keywords: housing cooperatives, Nordic model, community practices, neighbourhood conflict, householding, communal spaces

Highlights

- This chapter offers a sociocultural explanation as to why Nordic capital cities are less affected by the entry of short-term-accommodation rental platforms in the housing market.
- Communitarian movements in Nordic countries have shaped the social dynamics and resilience of Danish urban neighbourhoods.
- Housing cooperatives are miniature social laboratories of cohabitation as well as inclusive and participatory social practices.
- Nordic cultures of citizen engagement condition the response and collective action against P2P-rental platforms, because cooperative residents are constantly disciplined in the values and practices of belonging and participation.
- Resilient communities are forged by neighbours' engagement in conflicts and their efforts to resolve issues splitting or threatening the community.

Introduction

In this chapter I approach resilience as a problem-solving practice in neighbourhood communities, by discussing how peer-to-peer

*sgy.marktg@cbs.dk

DOI: 10.1079/9781789246605.0010

(P2P)-accommodation platform rentals have been embraced in Danish housing cooperatives. The spread of platform rentals brought about grave socio-economic effects for inner-city neighbourhoods throughout Europe, such as rising rent levels, conversion of residential stock into short-term commercial accommodation, evictions and housing speculation (Sans and Quaglieri, 2016; Freytag and Bauder, 2018; Cocola-Gant and Gago, 2019). In contrast, Nordic capital cities weathered the new market entrant without notable disruptions (Jokela and Minoia, 2020). This can be partially attributed to a policy context where the housing market is heavily regulated, but also to a cultural context where cooperative ideals and cooperative housing have been fostered for over a century. Instead of promoting the idea of Nordic exceptionalism, I offer a historical perspective shaping the dynamics of social cohesion in Danish neighbourhood communities and emergent, resilient ways to cope with P2P-platform rental.

Much of everyday urban life takes place in a spatial context shared with others: on crowded streets, public transport, parks and supermarkets and residential areas designed for cohabitation. The latter are the mundane (and perhaps overlooked) stages of community building. Our daily interactions with people living next door indicate the level of social cohesion in society. The organization of how communal areas are looked after reflects a neighbourhood's capability of collective action. Voluntary and compulsory joint activities may also have socializing effects on a broader scale, where neighbours develop a sense of care and trust for each other. It can thus be argued that the resilience capacity of a community is forged by mundane practices or 'doings' among neighbours, comprising encounters, helping out with banal chores or overcoming conflicts. As such, the everyday life of housing cooperatives can be seen as **integrative practice** (Schatzki, 1996), similar to farming, cooking or child-rearing, which is ordered by context-specific traditions and habits. Living with neighbours is conditioned by the so-called **teleoaffective structure**; a set of acceptable ends, conducts, routines and uses. For instance, in a Danish setting, the act of leaving a baby in a pram in a stairway or the backyard is considered an acceptable practice (young parents are even encouraged to let their babies sleep outside during the day). At the same time, it is not acceptable to be absent from compulsory maintenance activities or to throw parties in the garden without notifying neighbours. Hence, practices provide a palpable and fertile empirical context to study the interaction between the ideology and ways of communal living.

In order to shed new light on how social cohesion and resilience are built from the bottom, I will explore the evolution of Danish housing cooperatives and how this particular cohabitation form has shaped neighbourhood practices in the past and the present. The chapter is organized into three parts. First, it presents research approaches on platform rentals and neighbour conflicts and raises criticism with regards to an ahistorical and romanticizing perspective of local host communities before the advent of Airbnb. Second, it reviews the brief social history of the Danish cooperative housing phenomenon, with due attention to how conflicts and disputes have been handled in the past. Finally, it takes a practice theory perspective (Schatzki, 1996) on neighbourhood living, by revoking Beck's (2020) four-dimensional analysis of how community resilience is nurtured within everyday relations. By means of conclusions, I offer a discussion of implications for other cultural contexts.

Platform Rentals and Neighbourhood Conflicts

Just over a decade ago, the global market leader of short-term accommodation rental, Airbnb, put forward compelling claims about its business model. Most notably these claims presented a new and fairer economic alternative for peer transactions, which helped private hosts and entire host communities to generate supplementary income from renting out private property (Botsman and Capelin, 2016). Since then, an increasing body of research identified more controversial aspects of Airbnb's impacts (Dredge and Gyimóthy, 2015; Farmaki *et al.*, 2020). Critics provide wide-ranging evidence for Airbnb's role in 'touristification' and residential displacement processes (Guttentag, 2019), disruption in the urban housing markets (Wachsmuth and Weisler, 2018) and an overall deterioration of the quality of public space (Roelofsen and Minca, 2018). A growing body of empirical studies (Zervas *et al.*, 2017; Freytag and Bauder, 2018; Cocola-Gant and Gago, 2019; Ioannides *et al.*, 2019) indicates how platform rental businesses in housing markets are, in fact, contributing to a neoliberal economy 'on steroids', and instead of alleviating socio-spatial inequalities, they worsen labour precariousness and residential gentrification.

Researchers of the social impacts of platform accommodation rentals are unanimously concerned about their externalities and negative effects. Gottlieb (2013) highlighted how the business model of P2P rental disrupted the sense of community and idealistically harmonious everyday practices. The general image depicts a confrontational situation in which short-term rentals have exacerbated conflicts between neighbours. Permanent residents lament the loss of their private paradise as an effect of increased traffic and noise next door (Lorange and Oksanen, 2016). Trust among neighbours is broken when someone sublets their apartments on a regular basis to strangers who do not comply with the unwritten rules of the neighbourhood communities (Filippas and Horton, 2014). Once considered safe and protected, residential areas have turned into cosmopolitan attractions (Russo and Richards, 2016), making them more risky and prone to petty crimes and theft. The spaces of neighbourhood '*hygge*' [conviviality] are transformed into '*uhygge*' [creepiness or strangeness]. A few years ago, media coverage of Airbnb was dominated by extreme cases of abuse, such as illegal brothels (Christensen, 2016) or the complete damage of properties following orgies (Yuhas, 2015). While such instances are very rare, neighbours of P2P rentals echo these mediatized risk perceptions and fears, and consider themselves to be disempowered victims:

> It is also sad to experience the environment in your wonderful backyard being destroyed by two Airbnb-rentals. Only two years ago, children could play undisturbed in our backyard. Now we need to be afraid of old men that take photos of our small children in their summer clothes. It is simply 'uhyggeligt' [scary], and because of the guests, who never present any kind of ethics and moral, we do not dare to let the school children play alone in the garden. The guests make noise, drink, use drugs and leave food and other garbage (that the rats eat).
>
> (Lorange and Oksanen, 2016, p. 96)

Apart from the changing social dynamics associated with strangers in the neighbourhood, a more problematic issue is the increased economic cost inflicted on neighbours of P2P-rental property, especially in apartment buildings (Filippas and Horton, 2014). Short-term tenants identify themselves as guests exempted from duties and obligations and care little about investing in the neighbourhood, or protecting the values of the community (Gottlieb, 2013). When guests ravage a shared backyard or leave garbage in the hallway, the costs of cleaning up are carried by all neighbours. This is especially true for housing cooperatives where members share the responsibility of the operation and maintenance of the apartment building. In Helsinki, for instance, Jokela and Minoia (2020) identified the risk of Airbnb rentals breaking long-standing housing traditions. Although Airbnb promotes a sense of community and belonging in their brand communications to attract guests, the practices of 'being a local' (for instance, lounging in the communal backyard) do not extend to sharing the responsibilities with the permanent residents in the building.

Belonging to a neighbourhood assumes certain rights and duties towards community members. The appearance of P2P rental guests has now reignited debates around such fundamental roles (Lorange and Oksanen, 2016). The commercialization of private homes is free-riding on communal assets, including the maintenance of those. Yet, short-term guests staying in P2P rentals do not engage with the chores, leading to feelings of disquiet among permanent residents. In cohousing contexts, the most pressing issues identified by Andelsboligforeningers Fællesrepræsentation (ABF) [The Joint Council of Cooperative Housing Associations] include a fair taxation of rental revenues as well as the enforcement of public safety and protection of shared property (ABF, 2016). Such research insights are valid and indispensable to understand the various disruptive impacts of P2P rentals, yet they ignore the broader social and historical perspectives of cohabitation. Living with neighbours has never been entirely free of conflict or frictions. The tragedy of apartment-building commons (including free-riding and asocial behaviour) existed already before the dawn of P2P-rental platforms. Before holding Airbnb accountable for the disaggregation of perfectly harmonic communities, we should instead acknowledge that it has most likely intersected other extant social dynamics among residents. This necessitates a social historical perspective on those dynamics and changes in the decades preceding the touristification of residential areas.

Danish Housing Cooperatives

In contrast to most other countries around the world, communitarian ideals are inherent to Scandinavian culture. In particular, Denmark and Sweden have a long tradition of housing cooperatives and still promote an inclusive, affordable utopia and communal dream of cohabitation. Citizens of the Nordic countries embrace social cohesion and inclusion on a large scale, and strongly support the idea that affordable housing should be available to all members of society, so as to avoid the marginalization of less-affluent individuals (Bruun, 2011). For more than 100 years, Scandinavian communities experimented with managing private ownership collectively, through different forms of unions, agricultural cooperatives and homeowner associations. Cooperative forms of living have become especially popular in the wake of the demographic boom after World War II, and proved to be effective in reducing housing shortages and speculations in real estate investments. Today, one-third of the housing stock (approximately 100,000 flats) in Copenhagen is bound up in this form of property (Københavns Kommune, 2018). Housing cooperatives are seen as one of the key institutions beyond the family for developing strong social ties and to moderate economic, class and power differences. They can be regarded as moral hotbeds to develop the Nordic values of egalitarianism (advocating the removal of social and economic inequalities among people and the decentralization of power), where people practise democratic decision making and citizen control every day, for instance through voluntary participation in communal labour as well as sharing costs and profit (Bruun, 2011). As such, cooperative housing associations are important spaces of 'arm's length democracy', and they are run as independent, self-regulating societies with almost no interference from the state (Bruun, 2011).

However, the existence of Danish housing cooperatives has always been contingent on and shaped by national economic and housing policies. Initially, in the post-world-war era, Danish housing cooperatives were neither market oriented nor professionalized, compared with their private counterparts in the USA. The 1960s' Danish housing communities are regarded as an ideological revolt against the nuclear family; a 'commune dream' (Christensen and Kristensen, 1972) that emphasized shared economy, absence of property rights, gender and power equality, common child-rearing, voluntary division of labour, dissolution of one-to-one relationships and common outward-directed political and creative activities. They encapsulated an affordable utopia, 'creatively mix[ing] private and common dwellings to recreate a sense of community, while preserving a high degree of individual privacy' (Larsen 2019, p. 1355).

The growth of housing cooperatives peaked in the 1970s and 1980s, when the Danish government attempted to help out the domestic construction sector to reboot the economy amid an enduring crisis. The intervention entailed new legislation concerning cooperatives, which allowed housing associations to cover 80% of construction costs through state subsidized, interest-free loans (Erhvervsministeriet, 1980). This has kicked off a regular bonanza of new buildings available for cooperative tenure, much below market prices. Eventually, with the recovery of the Danish economy and the construction industry, the state support of new-built housing cooperatives was gradually phased out and terminated in 2004 (Erhvervs- og Byggestyrelsen, 2006; Andersen and Winther, 2010). In the wake of the past decades' neoliberal changes to the housing market, the Danish housing commons also experienced a massive commodification (Larsen and Hansen, 2015). Shifting property ownership and risky financial strategies (e.g. variable mortgage rates) until the 2008/09 global financial crisis brought many Danish housing cooperatives to the edge of insolvency. Rents rose beyond guaranteed levels, especially in Copenhagen's attractive neighbourhoods and the phenomenon of '*forældrekøb*' [wealthy parents buying cooperative property for their children as real estate investment] flourished. This excluded less-affluent tenants, who were originally the primary target group of this form of housing (Københavns Kommune, 2018). The marketization of housing cooperatives was criticized in media commentaries as 'greedy' and offensive to the traditional Danish principle of solidarity, but in fact, it was a result of removing state subsidies that were disrupting free market dynamics (Larsen, 2019). Nevertheless, the new economic realities have put the 'younger' housing cooperatives in greater Copenhagen under economic pressure, which subsequently has been balanced off through increased rental prices (Larsen and Hansen, 2015) and subletting initiatives (on a long-term basis).

Attitudes to P2P rentals in Danish housing cooperatives

Prior to 2016, subletting on short-term rental platforms had been restricted and prohibited in most housing cooperatives. Airnbnb's entrance in the Danish market stirred up an intense debate regarding changing those regulations, and thereby allowing more flexibility for economically vulnerable tenants to capitalize on their share of property. A survey conducted by ABF (2016) among 868 respondents revealed that 60% of cooperative tenants in Copenhagen supported a subsidiary legalization of short-term rental (compared with 25% in the rest of the country). Furthermore, 75% of respondents with previous experience as Airbnb guests were positive towards having their neighbours subletting their apartments to occasional guests. Although these attitudes are somewhat at odds with international research, claiming that tenants are ambivalent towards Airbnb in their neighbourhood, they are plausible in the light of the economic context discussed above. Subsequent to multi-layered discussions in associations, boards and at the national level, the ABF formulated a supplementary paragraph to standard statutes (ABF, 2019). This would allow tenants in individual associations to rent their homes for a short time for holiday and leisure purposes (ABF, 2019). The supplementary regulation determines: (i) the duration of the subletting period (a minimum of 3 days and a maximum of 3 weeks/year); (ii) the responsibilities of subletting members towards their neighbours in the property; and (iii) sanctions against those not complying with local conditions. Although the ABF does not intervene in how associations handle short-term rental locally, it offered extensive legal and practical advice on good conduct.

This showcases how the ABF, subsequent to decentralized discussions established a flexible framework for practice. The supplementary paragraph enabled local housing associations to determine whether or not they would allow subletting. It also facilitated working towards a win-win solution through continuous discussions and reflexive engagement with the ideals and concrete ways of co-living with short-term guests. However, for short-term guests, the implicit rules and meanings of integrative practices can be an unfamiliar terrain to navigate, and hence require more formal introduction. International guests, especially those from the USA, would find local routines and uses of common areas rather peculiar. As already discussed in the introduction, mundane practices (e.g. leaving prams containing babies or children playing unattended in a shared backyard) are deeply ingrained in a Nordic cultural context. Even banal and inconspicuous objects and doings in a neighbourhood are subject to community-specific rules and understandings. Residents are not only domesticated by a cooperative neighbourhood's specific dos and don'ts or 'order of life conditions' (Schatzki, 1996, p. 124), but also shape and co-create it. Accordingly, let us now take a closer look at neighbourhood practices in Danish housing cooperatives.

Practising neighbourhoods – neighbourhood practices

The performative lens allows us to see neighbourhoods as discursively and performatively constituted by residents through engaging in activities, meetings, discussions and struggles, which reproduce the value positions above. In fact, cooperative neighbourhoods are miniature subsistence systems in society, and their everyday practices are reminiscent of the notion of 'householding', described by Karl Polányi (1944). Householding, in its broadest sense, can be understood as joint subsistence and mutual exchange with members of extended households and neighbours. As such, householding performances can be regarded as experimental communitarian behaviour, aimed at sharing resources in more efficient and equitable ways. Beck (2020) has developed a conceptual framework that can be useful to understand householding practices in cohousing contexts. The framework cuts across four spatial dimensions: (i) value-oriented; (ii) organizational; (iii) relational; and (iv) physical.

The **value-oriented dimension** refers to an ideological vision of living together. How can we create principles of joint subsistence, where everyone can make the most of social living, but still keep some privacy? Despite cultural contexts that may differ and change over time, the ideals and tenets of cohabitation seem to revolve around stable and universal values. As Madden and Marcuse (2016) observed:

> For centuries, architects and utopians have pursued the idea that housing can provide the basis for a humane society. Such projects return again and again to the same themes: de-commodification, collective amenities, social spaces, democratic self-government, and engagement with the political and cultural life of residents.
>
> (Madden and Marcuse, 2016, p. 114)

These values, especially democratic self-government and active participation in the life of the community, which are central to Nordic countries, have also trickled down to the statutes of housing cooperatives. The way in which the vision is organized, institutionalized and implemented is expressed in the second dimension, namely the **organizational dimension**. For instance, all housing cooperatives have a board and members are represented on a one-member, one-vote basis. It is expected that members participate in association meetings as well as annual maintenance activities regularly, during which members collectively clean or repair common areas.

Democratic self-government used to be the lifeblood of cooperatives. However, this organizational practice is now under threat, as many apartments have changed proprietors (some even went into the hands of corporate investors) over the past few decades. In Copenhagen, 34% of the population has moved from one address to another in the past 5 years (Danmarks Statistik, 2021) and, in particular, the 25–34-year-old cohort is on the move. This mobility phenomenon may have contributed to a higher level of tolerance towards a frequent tenant/resident turnover in apartment buildings, but also to a lower propensity of engaging with the stranger next door. In other words, people take the community for granted. New members participate less frequently in meetings and communal chores, which may eventually break the cooperative relations that are at the core of Nordic cohabitation (Jokela and Minoia, 2020). As mentioned above, the ABF helps associations to establish the local organizational frameworks by providing financial, legal and social advice, but it is up to the individual associations to decide how to shape them to best comply with their members' needs and wishes. Such advice could include practical suggestions to comply with the supplementary regulation of short-term subletting on P2P platforms. The regulations explicitly state that it is the responsibility of the cooperative member to ensure that Airbnb guests comply with the house rules and other norms in the association (ABF, 2016, 2019). Examples include how the board and neighbours should be notified in advance of Airbnb visits or how to scrutinize potential guests (e.g. by requiring photographs or checking their reputation from platform host reviews) before accepting bookings.

The third, **relational dimension** is, therefore, of crucial importance to the understanding of cooperative living. Neighbourhoods are made up of 'doings', that is social interactions, each carrying the potential of reproducing or contesting the core values and meanings of living together. Making small talk with neighbours or sharing a drink in the backyard are mundane yet important acts of building trust and feelings of togetherness with the other residents. However, social meeting places for forging collectives are disappearing, even within the collectives, and so are meaningful social relationships between neighbours. Older tenants look back nostalgically to the bygone days, when life in housing associations took place mostly in communal areas and leisure was also a shared activity:

> We had a shared toilet, shower and laundry room. In the backyard we had a few square meters to grow herbs, sunbathe in peace and invite friends for a barbecue sausage or a good glass of wine.
>
> (Zander and Nathan, 2018)

Socializing thrives in communal spaces. Accordingly, the relational dimension is closely intertwined with the fourth, **physical dimension** of householding: semi-private, common and public areas. However, many of those social meeting places in Danish housing cooperatives have been quietly obliterated over the years or turned into commercial spaces. According to the CEO (chief executive officer) of Naboskab, Hanne Sophie Hjorth, the new interior design of communal housing is problematic, because it increasingly limits the spatial opportunities to meet and mingle with neighbours (Stender *et al.*, 2021). On the one hand, there is less space designated for common playgrounds or leisure activities, owing to the low commercial potential. On the other hand, residents also refrain from joint leisure activities. Instead of meeting people living next door in the backyard, many prefer to retreat to their private spaces. The recent construction of private balconies has accelerated these tendencies to isolate. The social downside of the balconies is that people increasingly shut themselves in. This, in turn, triggers disagreements between those who have balconies and those who do not.

The so-called balcony disputes are a case in point. In 2016, the City of Copenhagen received 2334 applications to install balconies, and more than 80% were approved (Stender *et al.*, 2021). Balconies make 'bad' neighbours; the number of legal cases increased subsequent to the instalments (Zander and Nathan, 2018). Some complained about the lack of privacy or daylight as neighbours' balconies cast a shadow on *their* apartments. Some even attempted to calculate the loss of market value due to their apartments losing their exclusive '*liebhaver*' (refers to unique, artistically designed properties or expensive condominiums with a high market value) qualities. Those who owned apartments with new balconies held on to their individual rights of construction approved by the municipality. The balcony disputes once again got residents riled up enough to engage in the debate and to look for alternative solutions. After an intensive and extended debate, a housing association in Valby (Skolehaven) opted out from building private balconies. Instead, they invested collectively in a 140 m^2 roof-terrace garden, (Zander and Nathan, 2018), where residents could grow various herbs and meet while gardening. The association Naboskab envisages that spending time with those living next door may increase empathy and prevent conflicts that may arise among neighbours. Such initiatives can be a new way of establishing the physical context of integrative, community-building practices and create: 'a sort of practical "lifebelt" for the modern human and an attempt to recreate the meaningful social relations that have been lost in the modern society in which we no longer know our neighbours' (Dansk Bygningsarv, 2016, p. 15).

Discussion: Theoretical Implications

The Danish housing cooperatives' handling of P2P rentals and the balcony disputes illustrate that cohesive neighbourhood practices are not necessarily abandoned by investments into (or marketization of) private property. Rather, such controversies are being addressed in the consolidated framework of public debates that characterize Nordic countries, such as through local and national association meetings or citizen councils. Neighbourhoods are thus not only a stage of unfolding householding practices, but also a platform where collective solutions to extant challenges and resilience are built (Halperin, 1998). Neighbourhoods undergo constant change and hence, housing cooperatives are the breeding grounds of resilience. Both old and new tenants, platform rental hosts and their neighbours constantly contest neighbourhood values, relations and spaces. Resilience can be thought of as being maintained along the circuit of practice (problematization, ritualization, instrumentalization) as described by Magaudda (2011) and Arsel and Bean (2013). First, the community *problematizes* and clarifies the issue that needs action. In both the case of the balcony disputes and the short-term rentals, neighbours gathered to discuss openly the nuisances they experienced, to address anxieties and loss of trust, and to find workable solutions. Second, the community established new legal and practical frameworks (*rituals*) to organize these activities in a way that was acceptable to the majority of tenants. In the case of subletting, bans and temporal limits were introduced, and prospective hosts received detailed instructions on how to embed guests in local settings. With the help of a cultural intermediary (the ABF), cooperatives *instrumentalized* good conduct and/

or restored good faith in neighbours renting out their property.

Performativity is, therefore, a valid approach to capture the resilience of communities facing disruptive events. For instance, during the Greek national economic crisis, we witnessed a sudden increase in ethnocentric consumption of domestic products (Lekakis, 2017). A similar collective awakening was reported in Sweden, where the citizens of Skåne successfully demonstrated and overturned the Arla group's decision of closing the regional dairy plant *Skånemejerier* (Humal and Schönström, 2016). Crises thus contribute to a heightened sense of communitarian consciousness and external threats may mobilize cooperative movements. Short-term P2P rentals advertised on online platforms can be regarded as an issue which constitutes a threat of a similar magnitude that puts pressure on many urban communities (Garay *et al.*, 2020).

Conclusion

In this chapter, I have argued that the everyday practices of Danish housing cooperatives are significant for building resilience in a community. Departing from the notion of the community as performatively constituted, I have provided a brief illustration of how neighbourhoods are enacted in residents' images or notions of cohesion as well as of shared values, practices and spaces. As the cases of P2P rentals and balcony disputes suggest, neighbourhood conflicts challenge but may also enhance societal cohesion, because they revive discussions about the rights and duties of community members. Housing cooperatives embarking on a capitalist trajectory (marketizing tenured property for profit) are reigniting citizen engagement for pragmatic, rather than ideological reasons.

The participatory culture of communitarian living is quite unique to the Nordic geographical context, and the transferability potential of similar practices to other regions may be somewhat limited. Nevertheless, the case of Danish housing cooperatives proves that adversity enhances the robustness of communities. Austerity has always spurred societal cohesion and communitarian behaviour, which also entailed experimenting with sharing resources in more efficient and equitable ways (Polányi, 1944). Cooperative movements in different sectors (agriculture, housing) are often triggered by economic or social crises, and inevitably revive discussions about the rights and duties of community members. The awakenings of communitarian consciousness may take place at different scales, ranging from national to regional and city levels. This opens up a fertile research avenue focusing on the connections between resilience identification and social consciousness in local communities. For instance, researchers could undertake empirical explorations of how neighbourhoods are being forged and consolidated when residents engage with issues splitting or threatening their community.

References

ABF (Andelsboligforeningers Fællesrepræsentation [The Joint Council of Cooperative Housing Associations]) (2016) Airbnb in Andelsboligforeningen. *Udeliv* (2), 4–8.

ABF (Andelsboligforeningers Fællesrepræsentation) (2019) Supplement to Statutes. Available at: https://www.abf-rep.dk/radgivning/abf-handbogen/ (accessed 1 July 2021).

Andersen, H.T. and Winther, L. (2010) Crisis in the resurgent city? The rise of Copenhagen. *International Journal of Urban and Regional Research* 34(3), 693–697.

Arsel, Z. and Bean, J. (2013) Taste regimes and market-mediated practice. *Journal of Consumer Research* 39(5), 899–917.

Beck (2020) What is co-housing? Developing a conceptual framework from the studies of Danish intergenerational co-housing. *Theory and Society* 37(1), 40–64. DOI: 10.1080/14036096.2019.1633398

Botsman, R. and Capelin, L. (2016) *Airbnb: Building a Revolutionary Travel Company*. Said Business School Case. University of Oxford, Oxford.

Bruun, M.H. (2011) Egalitarianism and community in Danish housing cooperatives. *Social Analysis* 55(2), 62–83.

Christensen, E. (2016) Airbnb-lejligheder brugt som bordel: der er svært at stoppe. Available at: https://nyheder.tv2.dk/samfund/2016-02-18-airbnb-lejligheder-brugt-som-bordel-der-er-svaert-at-stoppe (accessed 1 July 2021).

Christensen, S.K. and Kristensen, T.S. (1972) *Kollektiver i Danmark [Communes in Denmark]*. Borgen/Basis, Copenhagen.

Cocola-Gant, A. and Gago, A. (2019) Airbnb, buy-to-let investment and tourism-driven displacement: a case study in Lisbon. *Environment and Planning A: Economy and Space*. DOI: 10.1177/0308518X19869012

Danmarks Statistik (2021) Flytninger i Danmark. Available at: https://www.dst.dk/da/Statistik/emner/borgere/flytninger/flytninger-i-danmark (accessed 1 July 2021).

Dansk Bygningsarv (2016) *Fremtidens bofællesskaber* [Cohousing communities of the future]. Udlaendinge-, Integrations- og Boligministeriet, Copenhagen.

Dredge, D. and Gyimóthy, S. (2015) The collaborative economy and tourism: critical perspectives, questionable claims and silenced voices. *Tourism Recreation Research* 40(3), 286–302. DOI: 10.1080/02508281.2015.1086076

Erhvervsministeriet (1980) Cirkulære om offentlig støtte til opfaerelse og drift af andelsboliger. Available at: https://www.retsinformation.dk/eli/mt/1980/226 (accessed 1 July 2021).

Erhvervs- og Byggestyrelsen (2006) *Analyse af Andelsboligsektorens Rolle på Boligmarkedet [Analysis of the Cooperative Sector's Role in the Housing Market]*. Erhvervs- og Byggestyrelsen, Copenhagen.

Farmaki, A., Christou, P. and Saveriades, A. (2020) A Lefebvrian analysis of Airbnb space. *Annals of Tourism Research* 80: 102806.

Filippas, A. and Horton, J.J. (2014) The tragedy of your upstairs neighbours: when is the Airbnb negative externality internalized? Available at: http://papers.ssrn.com/sol3/papers.cfm?abstract_id=2443343 (accessed 1 July 2021).

Freytag, T. and Bauder, M. (2018) Bottom-up touristification and urban transformations in Paris. *Tourism Geographies* 20(3), 443–460. DOI: 10.1080/14616688.2018.1454504

Garay, L., Morales, S. and Wilson, J. (2020) Tweeting the right to the city: digital protest and resistance surrounding the Airbnb effect. *Scandinavian Journal of Hospitality and Tourism* 20(3), 246–267.

Gottlieb, C. (2013) Residential short-term rentals: should local governments regulate the 'industry'? *Planning and Environmental Law* 65(2), 4–9.

Guttentag, D. (2019) Progress on Airbnb: a literature review. *Journal of Hospitality and Tourism Technology* 10(4), 814–844. DOI: 10.1108/JHTT-08-2018-0075

Halperin, R.H. (1998) *Practicing Community: Class Culture and Power in an Urban Neighbourhood*. University of Texas Press, Austin, Texas.

Humal, K. and Schönström, J. (2016) Don't mess with my milk! The mystery of spectacular mundane consumption. Master's thesis, Lund University, Lund, Sweden. Available at: https://lup.lub.lu.se/luur/download?func=downloadFile&recordOId=8878691&fileOId=8878696 (accessed 1 July 2021).

Ioannides, D., Röslmaier, M. and van der Zee, E. (2019) Airbnb as an instigator of 'tourism bubble' expansion in Utrecht's Lombok neighbourhood. *Tourism Geographies* 21(5), 822–840.

Jokela, S. and Minoia, P. (2020) Nordic home-sharing utopia: a critical analysis of Airbnb in Helsinki. *Scandinavian Journal of Hospitality and Tourism* 20(3), 227–245. DOI: 10.1080/15022250.2020.1774412

Københavns Kommune [Municipality of Copenhagen] (2018) Analyse af tilgængeligheden på boligmarkedet i København. Available at: https://planinfo.erhvervsstyrelsen.dk/sites/default/files/media/publikation/analyse_af_tilgaengeligheden_paa_boligmarkedet_i_koebenhavn_koebenhavns_kommune_2018.pdf (accessed 1 July 2021).

Larsen, H.G. (2019) Three phases of Danish cohousing: tenure and the development of an alternative housing form. *Housing Studies* 34(8), 1349–1371. DOI: 10.1080/02673037.2019.1569599

Larsen, H.G. and Hansen, A.L. (2015) Commodifying Danish housing commons. *Geografiska Annaler* 97(3), 263–274. DOI: 10.1111/geob.12080

Lekakis, E.J. (2017) Economic nationalism and the cultural politics of consumption under austerity: the rise of ethnocentric consumption in Greece. *Journal of Consumer Culture* 17(2), 286–302.

Lorange, J. and Oksanen, N. (2016) An exploratory study of: peer-to-peer accommodation rentals in Copenhagen – a resident perspective. Master's thesis, Aalborg University, Aalborg, Denmark. Available at: https://projekter.aau.dk/projekter/files/239533963/Lorange_Oksanen_Online.pdf (accessed 1 July 2018).

Madden, D.J. and Marcuse, P. (2016) *In Defense of Housing: the Politics of Crisis*. Verso, London.

Magaudda, P. (2011) When materiality 'bites back': digital music consumption practices in the age of dematerialization. *Journal of Consumer Culture* 11(1), 15–36.

Polányi, K. (1944) *The Great Transformation: the Political and Economic Origins of Our Time*. Beacon Press, Boston, Massachusetts.
Roelofsen, M. and Minca, C. (2018) The Superhost: biopolitics, home and community in the Airbnb dream-world of global hospitality. *Geoforum* 91, 170–181. DOI :10.1016/j.geoforum.2018.02.021
Russo, A.P. and Richards, G. (eds) (2016) *Reinventing the Local in Tourism: Producing, Consuming and Negotiating Place*. Channel View Publications, Bristol, UK.
Sans, A.A. and Quaglieri, A. (2016) Unravelling Airbnb: urban perspectives from Barcelona. In: Russo, A.P. and Richards, G. (eds) *Reinventing the Local in Tourism: Producing, Consuming and Negotiating Place*. Channel View Publications, Bristol, UK, pp. 73–109.
Schatzki, T.R. (1996) *Social Practices: a Wittgensteinian Approach to Human Activity and the Social*. Cambridge University Press, New York.
Stender, M., Rudå, S. and Wolters, M. (2021) *Altaners sociale liv: Altaners betydning for boligliv, naboskab og byens fælles rum*. Report, BUILD, Aalborg University, Aalborg, Denmark. https://sbi.dk/Assets/Altaners-sociale-liv/Altanernes-sociale-liv.pdf (accessed 1 July 2021).
Wachsmuth, D. and Weisler, A. (2018) Airbnb and the rent gap: gentrification through the sharing economy. *Environment and Planning A: Economy and Space* 50(6), 1147–1170. DOI: 10.1177/0308518X18778038
Yuhas, A. (2015) Airbnb hosts return to find home trashed after 'drug-induced orgy'. *The Guardian*, 30 April. Available at: https://www.theguardian.com/technology/2015/apr/30/airbnb-calgary-home-trashed-drug-induced-orgy (accessed 1 July 2021).
Zander, I. and Nathan, I. (2018) Altaner skaber dårligt naboskab. Available at: https://www.dr.dk/nyheder/regionale/hovedstadsomraadet/virksomhed-altaner-skaber-daarligt-naboskab (accessed 1 July 2021).
Zervas, G., Proserpio, D. and Byers, J.W. (2017) The rise of the sharing economy: estimating the impact of Airbnb on the hotel industry. *Journal of Marketing Research* 54(5), 687–705.

11 Understanding the Airbnb Community and its Community Impact: the Use of Scenarios to Build Resilience

Jeroen Oskam*
Hotelschool The Hague, The Netherlands

Abstract

This conceptual chapter looks back at two scenario studies of the development of 'home-sharing', and in particular Airbnb, that were conducted in 2015 and 2018, respectively. These studies revolved around analyses of different interpretations and controversies that surrounded the emergent phenomenon: (i) the question, whether 'home-sharing' facilitated a new mode of consumption and, with that, the integration of tourists in resident communities; (ii) the empowering effects Airbnb claimed to have in resident communities; and (iii) the spatial distribution of tourism that 'home-sharing' would cause. These three debates are closely linked to the 'community' concept as it is used by Airbnb. The scenario studies gave insight, at an early stage, into the decisiveness of tourism growth in the eventual development of Airbnb. They also gave an early warning that without transparency local communities would be unable to impose their norms to control the phenomenon's undesirable impacts. The implication is that scenarios can help communities in strategic decision making.

Keywords: Airbnb, community impact, 'home-sharing', new urban tourism, 'overtourism', scenario planning

Highlights

- City authorities and resident communities failed to anticipate the disruptive impact of the exponential growth of 'home-sharing', as they generally misinterpreted it as an emancipatory and sustainable movement.
- Tourists 'living like a local' has not led to actual 'blending' into resident communities, but to a loss of services for residents and an alienation from their environment.
- The economic balance of 'home-sharing' for resident communities is negative, as generalized price increases outweigh individual profits from rental revenues.
- 'Home-sharing' has contributed to a further concentration of tourism pressure, rather than to its spreading.
- Scenario planning provides insights into unexpected events that cannot be extrapolated from current developments and can thus help communities build resilience by anticipating and managing disruptions.

Introduction

While reciprocal or altruistic home-sharing initiatives such as home swaps or Couchsurfing had been around for some time, profit-based

*J.Oskam@hotelschool.nl

DOI: 10.1079/9781789246605.0011

urban vacation rental platforms took off after the financial and economic crisis of 2008–2013. With the tailwind of economic recovery, these platforms showed an exponential development with triple-digit demand growth in cities like London, Barcelona and Amsterdam in 2015. The logics of platform markets pushed market leader Airbnb to marginalize competitors in a 'winner-takes-all' competition, turning the brand name into a generic denomination for monetary 'home-sharing'. The COVID crisis of 2020–2021 brought this evolution to a standstill, with little clarity about its possible future resurgence.

Both in the public and in the academic debate, the phenomenon was met with polarized interpretations (Martin, 2016; Buhalis *et al.*, 2020). As the vacation rental platform's discourse underscored concepts like 'sharing', 'community' and citizen empowerment, their growth initially resounded as a progressive and emancipatory movement (Airbnb, 2012; Gebbia, 2016). Tourism scholars embraced 'home-sharing' as a source of new authenticity for visitors blending in into cities' mundane neighbourhood life (Guttentag, 2015; Stors and Kagermeier, 2015; Paulauskaite *et al.*, 2017). Critics debunked the idealistic claims of the platforms, and also questioned the authenticity of the 'living like a local' experience (Morozov, 2013; Slee, 2015; Oskam and Boswijk, 2016; Murillo *et al.*, 2017; O'Regan and Chloe, 2017; Oskam, 2019, 2020a).

Although several studies have provided empirical evidence concerning the negative externalities of 'home-sharing', it is still hard to assess its long-term impact on resident communities, especially now that this impact has been suppressed by the sudden decrease of international and urban tourism since 2020 due to the COVID-19 pandemic. In 2015, a scenario panel comprising hotel professionals, real estate experts, tourism consultants and researchers conceptually examined the potential long-term effects of the phenomenon, with the objective of understanding its potential future impact on both the traditional tourism industry and on the city and its residents. This scenario-planning exercise was revisited by the author in 2018. In this chapter, I look back at those 'home-sharing' scenarios to see which of the underlying assumptions have shown to be accurate and which chain of events has actually evolved. This evaluation can show to what extent scenarios have been useful in supporting destination resilience.

These reflections only refer to 'home-sharing' in urban contexts. In coastal or rural destinations with a history of dependence on tourism, 'home-sharing' has generally added an additional sales channel to existing supply, with the advantage of attracting new customer segments and decreasing seasonality (Bijl *et al.*, 2017). In cities, 'home-sharing' has been responsible for the creation of additional supply – by transforming residential housing into tourist accommodation – rather than just acting as a distribution channel. Furthermore, the emergence of mass tourism in city destinations, precisely with the goal of blending into 'mundane city life' (Maitland and Newman, 2014), has turned out to profoundly alter that very city life. The 'blending in' of visitors has forced city residents and neighbourhoods into the role of tourist spectacle, an involuntary transformation that has contributed to the tensions that have surrounded recent developments in urban tourism (Novy, 2019; Novy and Colomb, 2020; Oskam, 2020a).

In this chapter I will discuss why tourist cities failed to adequately anticipate the impact 'home-sharing' would eventually have. Three different instances of the debate around 'home-sharing' show how the movement was initially embraced as a grassroots movement that would bring tourists and resident communities together and at the same time address inequality and sustainability issues. First, I will look at how research has corroborated or falsified these optimistic expectations. In the next section, I will analyse how Airbnb uses the concept of 'community'. Then, I will examine how resident communities, in hindsight, would have been more resilient to the changes associated with the emergence of 'home-sharing'. Finally, I will argue that scenarios can play an important role in building capacity to anticipate disruptive change.

The Community Impact of 'Home-sharing'

There are three contradictions to resolve in the predicted effects of 'home-sharing' on resident communities in cities. The first revolves around the social footprint of tourists 'living like a local'.

The second concerns the alienating and the empowering effects of 'sharing'. Finally, the third addresses the issue of spreading versus concentrating tourism. In what has been coined 'new urban tourism', visitors to a city aspire to behave as if they were not visitors, by avoiding traditional tourist sites and behaviour, and by participating in 'mundane city life'. This desire can be described as 'blending in', going 'off the beaten track' or 'living like a local'. The envisioned outcome is a situation in which 'divisions between tourism and everyday life in the city have blurred' (Maitland and Newman, 2014, p. 3). There are, however, two circumstances that impede a full immersion of visitors in city life. In the first place, they spend their time differently from residents and have different consumption patterns. In the second place, the division remains real as the 'local' becomes the primary aim of the visit; in other words, immersion is no longer possible if the number of other visitors exceeds a subjective threshold, in which case the destination is considered 'spoilt' or 'touristy'.

From the resident's perspective, the tension manifests itself when the visitor increases the demand for existing goods and services, and also requires different amenities, such as accommodation, prepared food rather than groceries, or tourist merchandise, thus making life more expensive while altering the available offer of goods and services. At the same time, residents are forced into the alienating role of involuntary actors on the stage of visitor immersion. Thus, residents establish a subjective threshold, beyond which they reinforce the division between themselves and visitors. They either begin to avoid places that attract visitors, or they might even plan new routes through their city. Some might end up leaving the city altogether (Quinn, 2007; Cócola Gant, 2015; Kamps, 2017; Oskam and Wiegerink, 2020).

Blurring divisions between locals and visitors is precisely one of the goals of 'home-sharing'. According to Airbnb, this has positive, empowering effects for city residents:

> Airbnb helps Barcelona families earn extra money to stay afloat during these difficult times by opening their homes to visitors,' said General Manager for Airbnb Spain and Portugal Jeroen Merchiers. 'In one year alone, thousands of local hosts helped travellers from around the world experience Barcelona more authentically and sustainably throughout the city's diverse neighborhoods.
>
> (Airbnb, 2013a)

> Sixty-five percent of Portland hosts said they have used Airbnb income to help afford staying [in] their homes. Hosting also helps Portland residents to pursue innovative careers and non-traditional forms of work. The study found 45 percent of hosts are selfemployed, freelancers or part-time workers, including 12 percent of hosts who have used Airbnb income to support themselves while launching a new business.
>
> (Airbnb, 2014)

At the same time, 'home-sharing' is an alienating force that turns mundane city life into a décor for tourist visits and residents into producers of commodified local ambiance. It can be argued that not only is this alienation not compensated by the empowering effects, but also the 'home-sharing' impact in fact undermines the economic advantages for residents. One of its effects for residents, for which there is overwhelming evidence, is that it drives up rents for residential housing (Samaan, 2015; Delgado-Medrano and Lyon, 2016; Lee, 2016; Schäfer and Braun, 2016; Horn and Merante, 2017; Barron *et al.*, 2018; Crommelin *et al.*, 2018; Wachsmuth and Weisler, 2018; Grisdale, 2019). In light of these increasing costs of living, the assertion that Airbnb helps people stay in their homes becomes questionable. One could wonder if without the platform, residents would even need the additional income: if 'home-sharing' has caused housing to become less affordable in the first place, offering one's house on Airbnb may even cease to be an act of choice, in which case the 'blurring' actually stands for a commodification of residents' personal space (Oskam, 2019).

Airbnb reports have claimed to cause tourism to spread outside the 'central hotel corridor' or the 'central tourist districts' (Airbnb, 2013b, c). This claim has found support in local policies that sought to alleviate pressure from congested tourist areas (Ecorys, 2018; Gemeente Den Haag, 2018; London First, 2018) and in academic studies, especially at the early stages of 'home-sharing' development (Guttentag, 2015; Stors and Kagermeier, 2015; Tussyadiah, 2015; Paulauskaite *et al.*, 2017). This idea of tourists leaving city centres to blend in with locals is crucial to the 'sharing' ideology as it would imply an actual shift in motives for and modes of consumption. Unfortunately, studies in different cities have demonstrated that Airbnb activity largely concentrates in traditional tourist areas (Arias Sans and Quaglieri, 2016; Gutiérrez *et al.*, 2017; Eugenio-Martin *et al.*, 2019), or at most expands to new gentrified tourist areas (Picascia *et al.*, 2017; Ioannides *et al.*, 2019). In London, despite people in peripheral and low-income neighbourhoods trying their luck on the platform, they were unsuccessful in luring tourists away from the central boroughs (Quattrone *et al.*, 2016). This 'centripetal' tendency in Airbnb demand has been shown to occur across 26 European cities (Oskam, 2020b).

The Airbnb Community

On the Airbnb website and in the words of the platform's spokespeople, 'community' is used in two different ways: it may refer to the 'local communities' in which 'home-sharing' takes place, but it mostly stands for the total of Airbnb users – hosts and guests. Their definition as a 'community' is a key part of the platform's marketing narrative. The underlying idea had been described previously by Airbnb's Global Head of Community Douglas Atkin (2004) as the 'culting' of a brand. Cult brands must not target the average customer, but those who feel different: 'Instead of trying not to alienate anyone, you must target the alienated and simultaneously separate your organization from the mainstream' (p. 18). In other words, this marketing strategy induces clients to exclusively identify with a single brand and, as 'true believers', to oppose or even feel hostility towards its alternatives. Just as cult brand Jetblue conveyed the message that it is not an airline (p. 39), Airbnb would become the anti-hotel, and Airbnb travellers would be anything but tourists.

Not only do Airbnb users differ from tourists and their accommodation from hotels: since they 'belong anywhere', their community supersedes all others. An analysis of promotion videos suggests that the Airbnb holiday enables community members to get in touch with their true selves, by bonding with others who, because of their idiosyncrasies, maybe display an equally untraditional, ambiguous relation with their home community (Oskam, 2019). However, at the same time Airbnb hosts find it important to underscore their local roots; their condition of 'born and raised' locals (Vermeulen, 2017) contests the accusation that 'home-sharing' harms communities and proves that the practice fits with locals more. For example, an advertisement in the campaign against the proposal to restrict the number of home-sharing rental days in San Francisco (Proposition F), invoked the local roots of its opponents as follows:

> San Franciscans aren't afraid of big challenges. We find ways to change the world. That's who we are. We always do it together. But I'm worried. I'm worried that Proposition F takes us in the wrong direction. It undermines the new short-term rental laws before they've even had a chance to work. Worse, Prop. F will block the city from being able to improve the law in the future. It goes too far. Vote no on Proposition F. It's just too extreme.
>
> (Vote No on Prop. F, 2015a)

Community Resilience

A community's resilience to tourism impacts denotes 'the ability of social, economic or ecological systems to recover from tourism-induced stress' (Tyrrell and Johnston, 2008, p. 16; Hall *et al.*, 2017), or specifically in urban contexts, 'to maintain or rapidly return to desired functions in the face of a disturbance, to adapt to change, and to quickly transform systems that limit current or future adaptive capacity' (Meerow *et al.*, 2016, p. 39). In the context of 'home-sharing', this means a resilience to alienation and tourist gentrification, an ability to absorb or reduce negative impacts such as nuisance, pollution or rent inflation, and to address the increased concentration of tourists in central locations.

As indicated above, these impacts are common in urban destinations but not generalized in all rural or coastal destinations; in tourist cities, the phenomenon affects the residential housing market, has year-round effects and contributes to the permanent discomfort of a peer-to-peer (P2P) 'tourist gaze'. Nevertheless, urban residential communities are not static, and 'home-sharing' or tourism growth in general are collectively but one driver in the alterations of their social fabric. Tourist gentrification may directly affect the lives of working-class residents (Cócola Gant *et al.*, 2020), or it may interfere with gentrified communities which, in turn, have replaced the original residents (Pinkster and Boterman, 2017).

Communities establish and enforce norms to protect themselves against the most obvious transgressions, such as noise, alcohol and drug abuse or crime, but also against disturbances of a desired social order, such as through subsidized housing or zoning plans. 'Sharing' platforms have fundamentally eroded this capability; ideologically, they have represented a movement for deregulation (Slee, 2015), advocating self-regulation based on review systems (Sundararajan, 2014, 2016). However, one of the limitations of this self-regulation is that it constitutes an alternative exclusively aimed at protecting the 'sharing' community against excesses from within. Airbnb users are certainly obliged to exhibit a certain politeness towards resident communities – for example not causing noise or other nuisance (Airbnb, 2019, 2020) – but this caution should be seen as a policy to prevent regulatory action that would jeopardize Airbnb supply. Airbnb does not recognize regulations that impede their users' entrepreneurial freedom (Griswold, 2015), or dismisses these as excessive, unjust or serving vested interests, to the absurd but repeated extreme that social movements such as the California Federation of Teachers or the leftist City Council of Barcelona under Ada Colau were accused of being biased towards the traditional hotel industry (Said, 2015; Vote No on Prop. F, 2015b; May, 2016).

The effectiveness of regulatory responses has been limited, for two different reasons. First, a strategic lack of transparency has prevented authorities from successfully enforcing policies, unless they have been able to negotiate registration systems with the platform, or they have opted for disruptive 'seek-and-destroy' policies to detect illegal rentals (Schneiderman, 2014; Lines, 2015; Cox and Slee, 2016; Herrera, 2016; Oskam, 2019). Second, these responses were, in general, initially hesitant and permissive, as they allowed for experimentation of what at the moment was considered an innovative and emancipatory movement (Gemeente Amsterdam, 2014).

Airbnb Scenarios

Scenario planning provides techniques to anticipate qualitatively different outcomes of current developments and uncertainties. This consideration of uncertain events makes it different from quantitative forecasting, 'usually constructed on the assumption that tomorrow's world will be much like today's' (Wack, 1985b, p. 73). Scenario planning was famously first used in a commercial setting at Royal Dutch Shell. In a market where forecast-based planning mostly foresaw how production and consumption would continue to grow, scenarios helped the company anticipate the volatility in the oil market of the 1970s (Wack, 1985a, b). Since then, scenario planning has become increasingly used, with the objective of developing strategic response capacity rather than making actual predictions of the future. It is, therefore, instrumental in building resilience to external disruptions (Bishop *et al.*, 2007; Hines, 2020). It is worth mentioning that scenario planning has been used in the context of destination level tourism in particular (Yeoman and McMahon-Beattie, 2005; Page *et al.*, 2006; Enger *et al.*, 2014).

In 2015, an expert panel discussed the drivers of the emergent 'home-sharing' phenomenon in Amsterdam, with the goal to understand potential future scenarios using the Global Business Network method (Schwartz, 2012). This first scenario study about 'home-sharing' in tourist cities drew conclusions different from the prevalent utopian interpretation of 'sharing' at that moment (Oskam and Boswijk, 2016), and concurred with the critical views of authors such as Morozov (2013), Cox (2015) and Slee (2015). The scenarios depicted four potential outcomes around two main uncertainties, plotted on two axes: (i) tourism growth in the destination; and (ii) the severity or permissiveness of local regulations. At the time of the study the explosive growth of tourism had just taken off in some cities, a driver that was expected to determine the force of the impact of 'home-sharing'. By contrast, more permissive rules would allow for innovation while strict regulation would stifle experimentation. This led to the scenario model shown in Fig. 11.1.

These scenarios saw a potential for experiential innovation with spin-offs for the wider hospitality and tourism sectors, but only if its

Demand is stable

	Repressive regulation	No regulation or Airbnb-friendly
Demand is stable	**1. Status quo** Second-tier tourist destinations that, apart from seasonality, attract constant visitor numbers. Closed markets with a limited number of well-known city hotels. Prices are stable except for the occasional event or congress when hotels maximize profits or – as the Spanish appropriately say – 'make their August'. It is also on these occasions when city authorities are somewhat more permissive towards peer-to-peer (P2P) rentals, which absorb the excess demand. Outside the peak moments there is no room for alternative accommodation. Hotels and authorities join forces to denounce and shut down illegal short-term rentals. Hotels follow trends to keep up with guest expectations.	**2. Experimentation** Second-tier destinations that are not traditionally considered tourist attractions but that want to refresh their image and stimulate tourist activity. The objective is to attract niche tourism. Non-traditional accommodation is seen as a way, or maybe even a condition, to become noticed by the desired type of visitors. Neither these visitors, nor the alternative accommodation, are considered a threat to the city's traditional hotel sector. On a limited scale – because of the niche character of tourism – innovative businesses and hybrid hotel-P2P concepts appear. There are no large-scale P2P 'multilistings'.
Booming demand	**3. Exclusivity** City destinations where tourism weariness inspire policies to decrease visitor numbers. Traditional hotels are protected and have an oligopolistic market control. Price hikes occur during events and high season. Low-cost tourists and students start avoiding the destination; affluent visitors, especially those from new markets, take their place. Hotels have no incentives to innovate and adapt to upscale markets and new visitors. Tourists drive up prices in restaurants and local retail. Illegal P2P rentals pop up like mushrooms to benefit from tourism; they get a 'hit-and-run' character as they are being shut down as soon as they are detected.	**4. Commercialization** City destinations that embrace booming tourism as an important economic sector. The growing size of the pie opens the market for new entrants. Innovative hotel types emerge, as well a full array of P2P accommodation: from basic to high-end luxury accommodation. Growth opportunities attract commercial parties, leading to a limited number of providers with 'multilistings' responsible for the larger part of the P2P offer. The suitability of a property for P2P rental becomes priced into rental fees, practically obliging residents to offer their houses as P2P accommodation. The result is gentrification of tourist areas and displacement of residents.

Booming demand

Fig. 11.1. Airbnb scenarios 2015. Four potential outcomes are shown based on two uncertainties plotted on two axes: (i) tourism growth in the destination; and (ii) the severity or permissiveness of local regulations. From Oskam and Boswijk, 2016.

development was accompanied by moderate growth of tourism; the hypothetical 'Experimentation' scenario was situated in Portland, Oregon (USA). Booming city tourism was expected to be disruptive if authorities would clamp down on Airbnb but even more if they would adopt a 'laissez-faire' policy. In this latter case, situated in Amsterdam, the price effects of tourism growth would be combined with residents' displacement because of housing market effects. This was illustrated – as is usual in scenario planning – with the extreme fictitious outcome of a popular neighbourhood turned into a timeshare where locals could only reside during the low seasons.

Despite the then favourable attitudes towards Airbnb of both authorities and the general public, this dramatic scenario was in hindsight hardly exaggerated; in particular, the increase of tourism pressure and weariness that was predicted to be associated with the growth of Airbnb became common in tourist cities, particularly in Europe. In the years that followed after the study, residents' protests (Colomb and Novy, 2017) and so-called 'overtourism' (Dodds and Butler, 2019; Milano *et al.*, 2019; Oskam, 2020c) became central topics in the political debate in Amsterdam and in many other cities. Visitor numbers to the city more than doubled from 2008 to 2018 (Oskam and Wiegerink, 2020); Airbnb overnight stays increased 474% from 19,000 in January 2015 to 107,000 1 year later; the yearly total of 736,000 booked nights increased to an estimated 1.4 million in 2017 (Bakker *et al.*, 2016, 2018). Even the absurdity of urban time-sharing has become a reality: both Amsterdam and Málaga have seen anecdotal evidence of so-called 'discontinuous rents', in which residents must give way to tourists in the summer months (Castillo, 2017; Roele, 2019).

These studies were part of a general shift in the public opinion around 'home-sharing', which by 2018 had led to a reversal of public policies and the imposition of rental restrictions. However, it could be argued that these changes came too late to recover from the 'tourist-induced stress' and to undo the disruptive impact of Airbnb. Against this background, I developed a new set of scenarios using the four 'archetypes' of the 'alternative futures' school: (i) continuation; (ii) collapse; (iii) discipline; and (iv) transformation (Dator, 2009). One of the central distinctions in these scenarios is that for local authorities to control the development of 'home-sharing', strict regulations would not be sufficient. Rather, an enforcement of regulations would also have to be facilitated by cities negotiating access to 'home-rental' data. Also, these scenarios nuanced the innovative image of Airbnb and foresaw a scenario of 'convergence' of 'home-sharing' with the traditional hotel industry, meaning that Airbnb was expected to move towards a hotel-like standardization while hotels would incorporate experiential innovations introduced by 'home-sharing' (Oskam, 2019).

Conclusion

The economic recovery after the crisis of 2008–2013 spurred a growth of urban tourism that affected resident communities and the use they made of functions and services in their cities: (i) it caused contagion, nuisance and pollution; (ii) it led to a proliferation of services targeting tourists replacing local services; (iii) it decreased the available housing stock, making it more expensive; and (iv) it even made residents re-plan their usual routes through the city. The emergence of 'home-sharing' was a catalyst in this process. These developments caught resident communities and local authorities off guard. The speed of change and an interpretation that in hindsight was too favourable caused their capacity to react and adjust, by imposing community norms, to be too limited and too late.

The main explanation of this favourable interpretation was the newspeak of 'sharing' marketing strategies that associated platforms such as Airbnb with Couchsurfing and similar idealistic movements rather than with profit-driven intermediation platforms, such as Booking.com or other online travel agencies. Thus, when Airbnb speaks of 'sharing', the company does not refer, as other people do, to the joint use of community assets (Belk, 2010), nor to 'the more efficient use of underutilized assets' (Botsman, 2010), but to 'commerce with the promise of human connection' (Gebbia, 2016). And when it offers tourists the proposition of 'living like a local', it does not mean that they will adopt the way of life of the local community, but rather that despite being tourists, they can be true to themselves, 'liberated from the compelling scripts of mass travel' (Oskam, 2019, p. 41).

'Community' is another sympathetic concept associated with 'sharing' and with doing things together. However, for Airbnb, it does not stand for the citizenry of a destination, but first of all for the platform's global customer base. 'Guests' do not seek contact with locals, as their interaction remains limited to the 'hosts'; all these people 'belong anywhere': their connection and cohesion supersede and deny other ties, such as those between hosts and their neighbours. The consequence is that, in the regulation debate, it is considered unfair, or simply incomprehensible, that a host's entrepreneurial freedom be subordinated to the interests of a community.

Scenarios provide a way to analyse the potential outcomes of disruptive or unexpected events and can thus support a community's resilience to external shocks. In 2015, a first scenario study indicated the importance of a regulatory response to 'home-sharing' from the side of resident communities, but it also foresaw the severity of its impact if accompanied by strong tourism growth. In 2018, a further study showed that for this response to be effective, it would require forcing the platform to transparency. These scenarios illustrate how the perception of 'home-sharing' has evolved, and hopefully have even contributed to that change in thinking. If so, this contribution is where our work as academics can be relevant for community resilience.

References

Airbnb (2012) Study Finds that Airbnb Hosts and Guests Have Major Positive Effect on City Economies. Available at: https://www.airbnb.nl/press/news/study-finds-that-airbnb-hosts-and-guests-have-major-positive-effect-on-city-economies (accessed 13 October 2015).

Airbnb (2013a) New Study: Airbnb Community Contributes $175 Million to Barcelona's Economy. Available at: https://www.airbnb.nl/press/news/new-study-airbnb-community-contributes-175-million-to-barcelonas-economy (accessed 13 October 2015).

Airbnb (2013b) New Study: Airbnb Community Contributes €185 million to Parisian Economy. Available at: https://www.airbnb.com/press/news/new-study-airbnb-community-contributes-185-million-to-parisian-economy (accessed 13 October 2015).

Airbnb (2013c) New Study: Airbnb Community Makes Amsterdam Economy Stronger. Available at: https://www.airbnb.nl/press/news/new-study-airbnb-community-makes-amsterdam-economy-stronger (accessed 13 October 2015).

Airbnb (2014) New Study: Airbnb Community Generates $61 Million in Economic Activity in Portland. Available at: https://www.airbnb.com/press/news/new-study-airbnb-community-generates-61-million-in-economic-activity-in-portland (accessed 13 October 2015).

Airbnb (2019) Building trust in our local communities. Available at: https://www.airbnb.com/resources/hosting-homes/a/building-trust-in-our-local-communities-126 (accessed 1 April 2020).

Airbnb (2020) Responsible hosting in the United States. Available at: https://www.airbnb.com/help/article/1376/responsible-hosting-in-the-united-states (accessed 1 April 2020).

Arias Sans, A. and Quaglieri, A. (2016) Unravelling airbnb: urban perspectives from Barcelona. *Built Environments and 'Glocalized' Spaces* 73, 209–228.

Atkin, D. (2004) *The Culting of Brands: When Customers Become True Believers*. Penguin, New York.

Bakker, D., Dessauvagie, M. and Oskam, J. (2016) *Airbnb: Impact and Outlook for Amsterdam 2015*. Colliers/Hotelschool The Hague, Amsterdam.

Bakker, D., Dessauvagie, M. and Oskam, J. (2018) *Airbnb in Nederland. De belangrijkste cijfers over 2017*. Colliers/Hotelschool The Hague, Amsterdam.

Barron, K., Kung, E. and Proserpio, D. (2018) The sharing economy and housing affordability: evidence from Airbnb. *SSRN Electronic Journal*. DOI: 10.2139/ssrn.3006832.

Belk, R. (2010) Sharing. *Journal of Consumer Research* 36, 715–734.

Bijl, J., Willemsen, C., Oskam, J. and Verhaak, L. (2017) *Airbnb aan de kust. De impact van Airbnb op de Nederlandse Noordzeekust*. Kenniscentrum Kusttoerisme, Vlissingen, the Netherlands.

Bishop, P., Hines, A. and Collins, T. (2007) The current state of scenario development: an overview of techniques. *Foresight* 9(1), 5–25.

Botsman, R. (2010) The case for collaborative consumption. TEDx Sydney. Available at: http://www.ted.com/talks/rachel_botsman_the_case_for_collaborative_consumption/ (accessed 19 October 2015).

Buhalis, D., Andreu, L. and Gnoth, J. (2020) The dark side of the sharing economy: balancing value co-creation and value co-destruction. *Psychology & Marketing* 37(5), 689–704.

Castillo, I.A. (2017) Los pisos turísticos inflan la burbuja de los alquileres con precios récord y poca oferta. La Opinión de Málaga. Available at: https://www.laopiniondemalaga.es/malaga/2017/11/27/pisos-turisticos-inflan-burbuja-alquileres/970558.html (accessed 22 April 2018).

Cócola Gant, A. (2015) Tourism and commercial gentrification. Paper presented at the RC21 International Conference on The Ideal City: Between Myth and Reality. Representations, Policies, Contradictions and Challenges for Tomorrow's Urban Life, 27–29 August 2015, Urbino, Italy. Available at: www.rc21.org/en/conferences/urbino2015/ (accessed 29 June 2018).

Cócola Gant, A., Gago, A. and Jover, J. (2020) Tourism, gentrification and neighbourhood change: an analytical framework – reflections from Southern European cities. In: Oskam, J. (ed.) *The Overtourism Debate: NIMBY, Nuisance, Commodification*. Emerald, Bingley, UK, pp. 121–135.

Colomb, C. and Novy, J. (2017) *Protest and Resistance in the Tourist City*. Routledge, London.

Cox, M. (2015) Inside Airbnb. Adding data to the debate. Available at: http://insideairbnb.com/ (accessed 31 October 2015).

Cox, M. and Slee, T. (2016) How Airbnb's data hid the facts in New York City. Inside Airbnb, New York. Available at: http://insideairbnb.com/reports/how-airbnbs-data-hid-the-facts-in-new-york-city.pdf (accessed 23 February 2021).

Crommelin, L., Troy, L., Martin, C. and Parkinson, S. (2018) Technological disruption in private housing markets: the case of Airbnb. *AHURI Final Report* 305. Australian Housing and Urban Research Institute Limited, Melbourne, Australia.

Dator, J. (2009) Alternative futures at the Manoa School. *Journal of Futures Studies* 14(2), 1–18.

Delgado-Medrano, H.M. and Lyon, K. (2016) *Short Changing New York City: the Impact of Airbnb on New York City's Housing Market*. BJH Advisors LLC, New York.

Dodds, R. and Butler, R. (2019) *Overtourism: Issues, Realities and Solutions*. De Gruyter, Berlin.

Ecorys (2018) *Toerisme in Amsterdam: vandaag en morgen*. Ecorys, Rotterdam.

Enger, A., Sandvik, K. and Iversen, E.K. (2014) Developing scenarios for the Norwegian travel industry 2025. *Journal of Tourism Futures* 1(1), 6–19.

Eugenio-Martin, J.L., Cazorla-Artiles, J.M. and González-Martel, C. (2019) On the determinants of Airbnb location and its spatial distribution. *Tourism Economics* 25(4), 1224–1244.

Gebbia, J. (2016) How Airbnb designs for trust. Ted.com. Available at: https://www.ted.com/talks/joe_gebbia_how_airbnb_designs_for_trust/transcript?language=en#t-751180 (accessed 6 April 2016).

Gemeente Amsterdam (2014) Memorandum of Understanding Gemeente Amsterdam and Airbnb. Binnenlands Bestuur. Available at: https://www.binnenlandsbestuur.nl/Uploads/2016/2/2014-12-airbnb-ireland-amsterdam-mou.pdf (accessed 18 December 2015).

Gemeente Den Haag (2018) *Toerisme, Werk(t) voor de Stad. Strategie Toerisme 2020–2025*. Gemeente Den Haag, The Hague, the Netherlands.

Grisdale, S. (2019) Displacement by disruption: short-term rentals and the political economy of 'belonging anywhere' in Toronto. *Urban Geography* 42(5), 654–680.

Griswold, A. (2015) Airbnb has a really convoluted process for being 'transparent' with New York City. *Quartz*, 15 December 2015. Available at: https://qz.com/571165/airbnb-is-sharing-more-data-but-in-a-really-convoluted-way/ (accessed 8 June 2018).

Gutiérrez, J., García-Palomares, J.C., Romanillos, G. and Salas-Olmedo, M.H. (2017) The eruption of Airbnb in tourist cities: comparing spatial patterns of hotels and peer-to-peer accommodation in Barcelona. *Tourism Management* 62, 278–291.

Guttentag, D. (2015) Airbnb: disruptive innovation and the rise of an informal tourism accommodation sector. *Current Issues in Tourism* 18(12), 1192–1217.

Hall, C.M., Prayag, G. and Amore, A. (2017) *Tourism and Resilience: Individual, Organisational and Destination Perspectives*. Channel View Publications, Bristol, UK.

Herrera, D.J. (2016) Federal judge gives initial backing to SF law that holds online rental platform companies accountable for illegal rentals. City Attorney of San Francisco, San Francisco, California. Available at: https://www.sfcityattorney.org/2016/11/08/federal-judge-gives-initial-backing-sf-law-holds-online-rental-platform-companies-accountable-illegal-rentals/ (accessed 14 May 2018).

Hines, A. (2020) Evolution of framework foresight. *Foresight* 22(5/6), 643–651.

Horn, K. and Merante, M. (2017) Is home sharing driving up rents? Evidence from Airbnb in Boston. *Journal of Housing Economics* 38, 14–24.

Ioannides, D., Röslmaier, M. and Van der Zee, E. (2019) Airbnb as an instigator of 'tourism bubble' expansion in Utrecht's Lombok neighbourhood. *Tourism Geographies* 21(5), 822–840.

Kamps, L. (2017) The tourist-resident encounter: impacts of tourism on the sense of place of Dutch students in inner-city Amsterdam. Master's thesis, Universiteit Utrecht, Utrecht, the Netherlands.

Lee, D. (2016) How Airbnb short-term rentals exacerbate Los Angeles's affordable housing crisis: analysis and policy recommendations. *Harvard Law & Policy Review* 10, 229–253.

Lines, G.E. (2015) Hej, Not Hej Da: Regulating Airbnb in the New Age of Arizona Vacation Rentals. *Arizona Law Review* 57, 1163.

London First (2018) *Tourist Information. Mapping the Local Value of International Visitors*. London First, London.

Maitland, R. and Newman, P. (2014) Developing world tourism cities. In: Maitland, R. and Newman, P. (eds) *World Tourism Cities*. Routledge, London, pp. 15–35.

Martin, C.J. (2016) The sharing economy: a pathway to sustainability or a nightmarish form of neoliberal capitalism? *Ecological Economics* 121, 149–159.

May, K. (2016) Airbnb and Homeaway hit back after fines in Barcelona. *Tnooz*. Available at: https://www.tnooz.com/article/airbnb-and-homeaway-hit-back-after-fines-in-barcelona/ (accessed 8 May 2018).

Meerow, S., Newell, J.P. and Stults, M. (2016) Defining urban resilience: a review. *Landscape and Urban Planning* 147, 38–49.

Milano, C., Cheer, J.M. and Novelli, M. (2019) *Overtourism: Excesses, Discontents and Measures in Travel and Tourism*. CAB International, Wallingford, UK.

Morozov, E. (2013) The 'sharing economy' undermines workers' rights. *Financial Times*, 14 October. Available at: https://www.ft.com/content/92c3021c-34c2-11e3-8148-00144feab7de (accessed 9 June 2020).

Murillo, D., Buckland, H. and Val, E. (2017) When the sharing economy becomes neoliberalism on steroids: unravelling the controversies. *Technological Forecasting and Social Change* 125, 66–76.

Novy, J. (2019) Urban tourism as a bone of contention: four explanatory hypotheses and a caveat. *International Journal of Tourism Cities* 5(1), 63–74.

Novy, J. and Colomb, C. (2020) Overdosed, underplanned or what? Making sense of urban tourism's 'politicisation from below'. In: Oskam, J. (ed.) *The Overtourism Debate: NIMBY, Nuisance, Commodification*. Emerald, Bingley, UK, pp. 75–94.

O'Regan, D.M. and Chloe, J. (2017) Airbnb: turning the collaborative economy into a collaborative society. In: Dredge, D. and Gyimóthy, S. (eds) *Collaborative Economy and Tourism*. Springer, Cham, Switzerland, pp. 153–168.

Oskam, J. (2019) *The Future of Airbnb and the 'Sharing Economy': The Collaborative Consumption of Our Cities*. Channel View Publications, Bristol, UK.

Oskam, J. (2020a) Commodification of the 'local' in urban tourism: the Airbnb contradiction. In: Oskam, J. (ed.) *The Overtourism Debate: NIMBY, Nuisance, Commodification*. Emerald, Bingley, UK, pp. 151–170.

Oskam, J. (2020b) Eiffel Tower and Big Ben, or 'off the beaten track'? Centripetal demand in Airbnb. *Hospitality & Society* 10(2), 127–155.

Oskam, J. (2020c) *The Overtourism Debate: NIMBY, Nuisance, Commodification*. Emerald, Bingley, UK.

Oskam, J. and Boswijk, A. (2016) Airbnb: the future of networked hospitality businesses. *Journal of Tourism Futures* 2(1), 22–42.

Oskam, J. and Wiegerink, K. (2020) The unhospitable city. Residents' reactions to tourism growth in Amsterdam. In: Oskam, J. (ed.) *The Overtourism Debate: NIMBY, Nuisance, Commodification*. Emerald, Bingley, UK, pp. 95–118.

Page, S., Yeoman, I., Munro, C., Connell, J. and Walker, L. (2006) A case study of best practice – Visit Scotland's prepared response to an influenza pandemic. *Tourism Management* 27(3), 361–393.

Paulauskaite, D., Powell, R., Coca-Stefaniak, J.A. and Morrison, A.M. (2017) Living like a local: authentic tourism experiences and the sharing economy. *International Journal of Tourism Research* 19(6), 619–628.

Picascia, S., Romano, A. and Teobaldi, M. (2017) The airification of cities. Making sense of the impact of peer to peer short term letting on urban functions and economy. In: *Proceedings of the Annual Congress of the Association of European Schools of Planning (AESOP '17). Spaces of Dialog for Places of Dignity: Fostering the European Dimension of Planning*, 11–14 July 2017, Lisbon, Portugal. AESOP, Lisbon, pp. 2212–2223.

Pinkster, F.M. and Boterman, W.R. (2017) When the spell is broken: gentrification, urban tourism and privileged discontent in the Amsterdam canal district. *Cultural Geographies* 24(3), 457–472.

Quattrone, G., Proserpio, D., Quercia, D., Capra, L. and Musolesi, M. (2016) Who benefits from the sharing economy of Airbnb? In: *Proceedings of the 25th International World Wide Web Conference*, Montreal, Canada, 11–15 April 2016. International World Wide Web Conferences Steering Committee, Montreal, Canada, pp. 1385–1394.

Quinn, B. (2007) Performing tourism. Venetian residents in focus. *Annals of Tourism Research* 34(2), 458–476.

Roele, J. (2019) Vrouw gedwongen om haar woning te verhuren via Airbnb. Het Parool. Available at: https://www.parool.nl/amsterdam/vrouw-gedwongen-om-haar-woning-te-verhuren-via-airbnb~bbca39d5/ (accessed 19 November 2019).

Said, C. (2015) Prop. F: S.F. voters reject measure to restrict Airbnb rentals. SF Gate. Available at: https://www.sfgate.com/bayarea/article/Prop-F-Measure-to-restrict-Airbnb-rentals-6609176.php (accessed 15 May 2018).

Samaan, R. (2015) *Airbnb, Rising Rent and the Housing Crisis in Los Angeles*. LAANE [Los Angeles Alliance for a New Economy], Los Angeles, California.

Schäfer, P. and Braun, N. (2016) Misuse through short-term rentals on the Berlin housing market. *International Journal of Housing Markets and Analysis* 9(2), 287–311.

Schneiderman, E. (2014) *Airbnb in the City*. Office of the New York State Attorney General Eric T. Schneiderman, New York.

Schwartz, P. (2012) *The Art of the Long View: Planning for the Future in an Uncertain World*. Currency Press, Redfern, New South Wales, Australia.

Slee, T. (2015) *What's Yours is Mine. Against the Sharing Economy*. OR Books, New York.

Stors, N. and Kagermeier, A. (2015) Motives for using Airbnb in metropolitan tourism—why do people sleep in the bed of a stranger? *Regions Magazine* 299(1), 17–19.

Sundararajan, A. (2014) Trusting the 'sharing economy' to regulate itself. *The New York Times*, 3 March. Available at: https://economix.blogs.nytimes.com/2014/03/03/trusting-the-sharing-economy-to-regulate-itself/ (accessed 3 June 2018).

Sundararajan, A. (2016) *The Sharing Economy: the End of Employment and the Rise of the Crowd-Based Capitalism*. MIT Press, Cambridge, Massachusetts.

Tussyadiah, I.P. (2015) *An Exploratory Study on Drivers and Deterrents of Collaborative Consumption in Travel. Information and Communication Technologies in Tourism*. Springer, Cham, Switzerland.

Tyrrell, T.J. and Johnston, R.J. (2008) Tourism sustainability, resiliency and dynamics: towards a more comprehensive perspective. *Tourism and Hospitality Research* 8(1), 14–24.

Vermeulen, R. (2017) Airbnb-hosts zijn juist geweldige ambassadeurs. *Het Parool*, 29 July. Available at: https://www.parool.nl/columns-opinie/airbnb-hosts-zijn-juist-geweldige-ambassadeurs~b5beebd6/ (accessed 7 July 2019).

Vote No on Prop. F (2015a) Lt. Gov. Gavin Newsom says Vote No on Prop. F. Available at: https://archive.org/details/SFPropF2015ads/nof_Lt._Governor_Gavin_Newsom_says_Vote_No_on_Prop._F.mp4 (accessed 27 March 2020).

Vote No on Prop. F (2015b) What Will Happen if Prop. F Passes. Available at: https://archive.org/details/SFPropF2015ads/nof_What_Will_Happen_if_Prop_F_Passes.mp4 (accessed 27 March 2020).

Wachsmuth, D. and Weisler, A. (2018) Airbnb and the rent gap: gentrification through the sharing economy. *Environment and Planning A: Economy and Space* 50(6), 1147–1170.

Wack, P. (1985a) Scenarios: shooting the rapids. How medium-term analysis illuminated the power of scenarios for shell management. *Harvard Business Review* 63(November–December), 139–150.

Wack, P. (1985b) Scenarios: uncharted waters ahead. *Harvard Business Review* 63(September–October), 73–89.

Yeoman, I. and McMahon-Beattie, U. (2005) Developing a scenario planning process using a blank piece of paper. *Tourism and Hospitality Research* 5(3), 273–285.

Index

Note: Page numbers in **bold** type refer to **figures** and Page numbers in *italic* type refer to *tables*

CABI – who we are and what we do